Have Blue and the F-117A: Evolution of the "Stealth Fighter"

Have Blue and the F-117A: Evolution of the "Stealth Fighter"

David C. Aronstein
Albert C. Piccirillo
ANSER
Arlington, Virginia

American Institute of Aeronautics and Astronautics, Inc.
1801 Alexander Bell Drive
Reston, Virginia 20191

Publishers since 1930

American Institute of Aeronautics and Astronautics, Inc., Reston, Virginia

Library of Congress Cataloging-in-Publication Data

Aronstein, David C.
Have Blue and the F-117A : evolution of the "Stealth Fighter" /
David C. Aronstein, Albert C. Piccirillo.
p. cm.
Includes bibliographical references (p.) and index.
1. F-117 (Jet fighter plane) —Design and construction. 2. United States. Air Force—
Weapons systems.
I. Piccirillo, Albert C. II. Title.
TL671.2.A624 1997 623.7'464—dc21 97-15440
ISBN 1-56347-245-7

Cover design by Sara Bluestone

Second printing.

Acknowledgments

The authors would like to thank Lt Gen George K. Muellner, U.S. Air Force, Deputy Assistant Secretary of the Air Force (Acquisition), for his interest, encouragement and strong support, without which the present work would not have been undertaken. Others who supported the authors in this endeavor include Dr. Paul G. Kaminski, Under Secretary of Defense for Acquisition and Technology and former Director of the U.S. Air Force low observables project office in the Pentagon; Lt Gen Richard M. Scofield, U.S. Air Force, recently retired Commander of the U.S. Air Force Aeronautical Systems Center at Wright–Patterson Air Force Base and former Senior Trend Program Manager; and Richard P. Hallion, Historian of the Air Force.

We would like to express our appreciation to Col Neil Kacena, U.S. Air Force, Director of Electronic and Special Programs within the Office of the Assistant Secretary of the Air Force (Acquisition), and to Maj John Fagnant, U.S. Air Force, the F-117A Program Element Monitor, for providing valuable input as well as access to Have Blue and Senior Trend archival materials, and for coordinating and facilitating the security review of this book. We would also like to thank Lt Col Randy A. Smith, U.S. Air Force, Joint Strike Fighter Flight Systems Integrated Product Team Lead, formerly of the 2762nd Logistics Squadron (Special), Detachment 6, Norton Air Force Base, for his support in pursuing this task as well as his input on various aspects of special projects management and logistics; and Wayne M. O'Connor of the Development Planning Directorate at the Aeronautical Systems Center, Wright–Patterson Air Force Base, for his strong support in the technical review of this document.

The following participants in the Have Blue and Senior Trend programs were kind enough to review our work or provide valuable input, which we have incorporated. The listing is based on organizational affiliation during the period covered in this book, and is roughly chronological by date of initial assignment within each organization.

Defense Advanced Research Projects Agency: Ken Perko, who initiated and managed the early Experimental Survivable Testbed studies.

Aeronautical Systems Division (ASD), Wright–Patterson Air Force Base: Herbert (Skip) Hickey, Flying Qualities Engineer for Experimental Survivable Testbed/Have Blue, later Chief Engineer for Senior Trend; Col John K. Twigg, U.S. Air Force (Ret), Program Manager for Have Blue, later the Director of the Classified Aeronautical Systems Program Office at ASD; Lt Col Peter Knauth, U.S. Air Force (Ret), Security Manager for the Classified Aeronautical Systems Program Office; Lynn Brucker, Avionics Engineer for survivability and system requirements analyses; and Joseph Reiman, Aerospace Engineer for signature and low observables supportability.

Lockheed Advanced Development Projects: Ed Martin, who played a key role in initiating Lockheed's Experimental Survivable Testbed program; Richard Scherrer, Head of the Preliminary Design Team for Experimental Survivable Testbed/Have Blue; Denys D. Overholser, Lockheed's primary expert in radar cross section computation and measurement during Experimental Survivable Testbed/Have Blue; Alan C. Brown, Chief Engineer for Have Blue and first Program Manager for the F-117A; and Sherman N. Mullin, F-117A Program Manager from 1982 to 1985, and President and General Manager of the Lockheed Skunk Works from 1991 to 1994.

Air Force Low Observables Project Office, the Pentagon (U.S. Air Force/RDPJ): Col Jerry W. Baber, U.S. Air Force (Ret), Contracting Officer from the initiation of the low observables effort until 1979, then Business Manager at the Classified Aeronautical Systems Program Office at ASD, and presently the Director of Contracts at the Lockheed Martin Skunk Works; and Col Michael D. Marks, U.S. Air Force (Ret), Senior Trend Program Monitor at RDPJ from 1981 through 1983, Director of the office from 1987 to 1989 (by which time the office symbol was SAF/AQL), and presently Vice President and General Manager of the F-15 Program at McDonnell Douglas.

4450th Tactical Group: Brig Gen James S. Allen, U.S. Air Force (Ret), Commander of the 4450th Tactical Group during the critical period prior to and including Initial Operational Capability of the F-117A.

Senior Trend Combined Test Force: Col Ted F. Bowlds, U.S. Air Force, Flight Test Engineer for avionics, and presently the Commander of the Air Force's Rome Laboratory; Lt Col Stanley P. Siefke, U.S. Air Force, Flight Test Engineer for performance and weapons separation; and Lt Col Charles Pinney, U.S. Air Force, Flight Test Engineer for avionics and weapons integration.

Special thanks go to Lockheed Martin Corporation for generously providing extensive photographic coverage of the Have Blue and F-117A aircraft.

Finally, we would like to acknowledge the vital role that Heather Brennan of the AIAA Book Program played as editor for this effort. Her true professionalism really made a difference.

Disclaimer

The opinions presented in this book are those of the authors and do not represent official Department of Defense or U.S. Air Force opinions or policy.

About the Authors

David C. Aronstein is a principal aerospace engineer at ANSER in Arlington, Virginia, where he provides senior technical support to several aerospace programs including the Joint Strike Fighter and the F-22 air superiority fighter. Previously, he was employed at the Boeing Commercial Airplane Company on the 777 and High Speed Commercial Transport programs in subsonic and supersonic aerodynamic design and analysis. Dr. Aronstein also has performed aircraft structural analysis and experimental fluid dynamics research related to the internal cooling flows in rotating turbine engine components. He holds a bachelor of science degree in mechanical and aerospace engineering with highest honors from Princeton University, a master of science degree in aeronautics and astronautics from Stanford University, and a doctor of philosophy degree in aeronautics and astronautics from the University of Washington.

Albert C. Piccirillo, Colonel, U.S. Air Force (Ret), is a principal aerospace engineer and manager of the Joint Technology Division at ANSER where he is responsible for leading technical and analytical efforts on advanced weapons systems including the Joint Strike Fighter. He previously was with Planning Research Corporation where he headed the C^3 Systems Department and managed efforts with both United States and foreign clients including Mitsubishi Corporation. As an Air Force colonel, he served as the Advanced Tactical Fighter Program manager and headed the source selection, which resulted in the YF-22 and YF-23 air superiority prototypes. Previously, he was chief of the Tactical Systems Division in Air Force Studies and Analyses responsible for analyzing advanced aircraft, weapons, and C^3I concepts. Prior to that, he was involved in a number of advanced weapons projects including development of conventional cruise missiles. At the Air Force Flight Test Center, he led an engineering team responsible for flight test of unmanned air vehicles. Earlier in his U.S. Air Force career, he served in operational and staff assignments throughout the world including a combat tour in Southeast Asia as an F-4 pilot. Mr. Piccirillo holds a bachelor of science degree in aeronautical engineering from Pennsylvania State University and a master of science degree in aerospace engineering from the Air Force Institute of Technology.

Table of Contents

List of Figures

List of Tables

List of Acronyms and Symbols

α	angle of attack (also referred to as alpha, AOA)
A/C	aircraft
ACC	Air Combat Command: a single command that combines the roles of the earlier Tactical Air Command (TAC) and Strategic Air Command (SAC), except for strategic ballistic missile forces
ACM	Advanced Cruise Missile
ACTD	advanced concept technology demonstration: a current category of technology demonstration project that need not be a part of a formal acquisition program
ADP	Advanced Development Projects division of Lockheed, the "Skunk Works"; sometimes abbreviated LADP
AFAL	Air Force Avionics Laboratory, which sponsored much of the early work in radar cross section prediction and reduction
AFFTC	Air Force Flight Test Center
AFLC	Air Force Logistics Command
AFOSI	Air Force Office of Special Investigations
AFOTEC	Air Force Operational Test and Evaluation Center (formerly AFTEC)
AFSC	Air Force Systems Command
AFTEC	Air Force Test and Evaluation Center (now AFOTEC)
AMPS	automated mission planning system
AMST	advanced medium STOL transport
AOA	angle of attack
APB	acquisition program baseline
ARI	aileron–rudder interconnect
ARM	anti-radiation missile
ASD	Aeronautical Systems Division (now Aeronautical Systems Center)
ATA	Advanced Technology Aircraft, original designation for the F-117A
ATB	Advanced Technology Bomber; became the Northrop B-2
ATF	Advanced Tactical Fighter
ATS	air-to-surface
β	yaw angle
BDA	bomb damage assessment
BES	budget estimate submission
BSAX	battlefield surveillance aircraft, experimental
CAD	computer aided design
CAM	computer aided manufacturing
CC	commander

CE/D	concept exploration/definition
CIA	Central Intelligence Agency
Cn_β	Variation of yawing moment coefficient with yaw angle; the primary measure of directional stability
COEA	cost and operational effectiveness analysis
CPFF	cost plus fixed fee (contract type), usually has a ceiling
CPIF	cost plus incentive fee (contract type)
CTF	Combined Test Force, referring to the Senior Trend flight test organization (TAC, AFSC, and contractors); JTF, or Joint Test Force, also used
CV	vice commander
CY	calendar year
D&F	Determination and Findings
DAB	Defense Acquisition Board
DARPA	Defense Advanced Research Projects Agency (now ARPA, the Advanced Research Projects Agency)
DCAS	Defense Contract Audit Service
DCS	Deputy Chief of Staff
DCS/R&D	Deputy Chief of Staff for Research and Development
DDR&E	DoD Director of Defense, Research, and Engineering
DEPSECDEF	Deputy Secretary of Defense
DLIR	downward-looking infrared
DO	Director of Operations
DoD	Department of Defense
DSA	Delegation of Special Authority
DSB	Defense Science Board
DT&E	developmental test and evaluation: testing focused on finding and fixing problems, refining, and improving a system
DX	defense expedite: a designation for a high-priority program
ECP	engineering change proposal
ECS	environment control system
EMD	engineering and manufacturing development: the phase in which highly representative prototypes or initial production units of a weapon system are developed, built, tested, and evaluated (formerly full-scale development and full-scale engineering development; See also FSED
F&O	field and organizational (maintenance)
FBW	fly-by-wire
FCD	full-capability development
FFP	firm fixed price (contract type)
FLIR	forward-looking infrared
FMC	full mission capable
FMC/S	full mission capable, Stealth

FPI	fixed price incentive (contract type)
FPIF	fixed price incentive firm
FSED	full-scale engineering development, a term that apparently came into use for a short period in the early 1980s for what is now called engineering and manufacturing development (full-scale development was the term in use at the start of F-117A development) see also EMD
Fw	Focke-Wulf
FY	fiscal year
FYDP	five year defense plan (now denotes future year defense plan)
GFAE	Government Furnished Aerospace Equipment
GFE	Government Furnished Equipment
GFP	Government Furnished Property
GO	Geometric Optics
GPS	Global Positioning System
GTD	Geometric Theory of Diffraction
HAC	House Appropriations Committee
HARM	High Speed Anti-Radiation Missile
HASC	House Armed Services Committee
HQ	Headquarters
IADS	Integrated Air Defense Systems (Soviet)
ILDC	Incremental Length Diffraction Coefficient
ILEDC	Incremental Length Edge Diffraction Coefficient
INS	Inertial Navigation System
IOC	initial operational capability
IOT&E	initial operational test and evaluation; see also DT&E, OT&E
IRADS	Infrared Acquisition and Designation System: a system unique to the F-117A that consists of a FLIR and a DLIR with associated hardware and software to perform self-contained, passive, night time recognition and designation of targets for the F-117A's guided weapons
JAST	Joint Advanced Strike Technology (now the Joint Strike Fighter program)
JROC	Joint Requirements Oversight Council
JSF	Joint Strike Fighter
JTF	joint test force; see also CTF
JTT	joint test team
KIAS	knots indicated air speed
LADC	Lockheed Advanced Development Company. See also LADP

LADP	Lockheed Advanced Development Projects, the "Skunk Works"; renamed Lockheed Advanced Development Company (LADC) in 1990 (sometimes referred to as ADP)
LEX	leading-edge extension
LGB	laser-guided bomb
LIOC	limited initial operational capability
LO	low observable
LPI	low probability of intercept
LRU	line replaceable unit
LTV	Ling–Temco–Vought
LWF	Lightweight Fighter
MC	mission capable
MFD	multi function displays
MILCON	military construction
MIL SPEC	military specification
MIP	maintainability improvement program
MMH/FH	maintenance man hours per flight hour
MNS	mission needs statement
NACA	National Advisory Committee on Aeronautics; predecessor of NASA
NASA	National Aeronautics and Space Administration
NAVPRO	Navy plant representative
OAT	outside ambient temperature
OCIP	offensive capability improvement program
OMB	Office of Management and Budget
OPEC	Organization of Petroleum Exporting Countries
ORD	operational requirements document
OSD	Office of Secretary of Defense
OT&E	operational test and evaluation: testing focused on evaluating the performance of a system or its capability to perform its intended mission
P3I	pre-planned product improvement
PAA	primary aircraft authorized (in a squadron)
PBD	program budget decision
PD & RR	program definition and risk reduction
PGM	precision guided munition
PMD	program management directive
PMRT	program management responsibility transfer
PO	Physical Optics
POM	program objective memorandum
PTD	Physical Theory of Diffraction

QC	quality control
QRC	quick reaction capability
R&D	research and development
R&E	research and engineering
R&M	reliability and maintainability
RAM	radar absorbent material
RATSCAT	Air Force Radar Target Scatter facility in New Mexico
RCS	radar cross section
RDPJ	Strategic reconnaissance office at Air Force Headquarters in the Pentagon where the Air Force low observables program was managed in its early stages
RDQ-LO	Air Force Special Projects Office at the Pentagon (formerly RDPJ, now SAF/AQL)
RDT&E	research, development, test, and evaluation (budget category)
RIP	reliability improvement program
RPV	remotely piloted vehicle
RTO	responsible test officer
SAC	Senate Appropriations Committee
SAC	Strategic Air Command
SAF/AQ	Assistant Secretary of the Air Force (Acquisition)
SAF/AQL	Office for special projects within SAF/AQ: performs the role formerly filled by the RDPJ office; see also RDPJ, RDQ-LO
SAM	surface-to-air missile
SAR	special access required
SASC	Senate Armed Services Committee
SE	support equipment
SECAF	Secretary of the Air Force
SECDEF	Secretary of Defense
SPN-GEAN	inertial navigation unit used initially on the Boeing B-52 and later on the F-117A
SPO	system program office
SRU	shop replaceable units
STAR	system threat analysis report
STOL	short takeoff and landing
T&E	test and evaluation
TAC	Tactical Air Command
TAC/DO	Director of Operations, Tactical Air Command
TEMP	Test and Evaluation Master Plan
TEU	trailing-edge up
TFW	tactical fighter wing
TG	tactical group

TSSAM	Tri-Service Standoff Attack Missile
TY	then-year
USAAF	United States Army Air Force: the Air Force before it became an independent service
U.S. Air Force/RD	Office of the Deputy Chief of Staff for Research and Development (DCS/R&D) at Air Force Headquarters
U.S. Air Force/RDPJ	early designation for the strategic reconnaissance office where the low observables effort was initially "hidden" within U.S. Air Force/RD; sometimes referred to as AF/RDPJ or RDPJ
USD	Under Secretary of Defense
USD/A&T	Under Secretary of Defense for Acquisition and Technology
UTE	utilization rate: sorties per aircraft per 24 h period
V/STOL	vertical/short takeoff and landing
VLO	very low observables
WRSK	war readiness spares kit
WSCS	Weapon System Computational Subsystem
WST	Weapon System Trainer (simulator)
XST	Experimental Survivable Testbed: early designation for Have Blue

Chapter 1

Introduction

Overview

This book describes the successful approach used by the U.S. Air Force to develop and field the first true very low radar signature, low observable (stealth) strike aircraft. That aircraft, the Lockheed F-117A, had its genesis in 1974, in the wake of the Southeast Asia and 1973 Arab/Israeli conflicts. These conflicts had demonstrated the potent capabilities of Soviet radar-based air defense systems. As a result of these experiences, the Defense Advanced Research Projects Agency (DARPA) initiated conceptual studies to determine whether it was possible to develop a manned aircraft with a sufficiently low radar cross section to defeat such systems.

The DARPA studies continued through the summer of 1975, by which time two of the participants — Lockheed and Northrop — appeared to have achieved breakthroughs in the ability to design low observable aircraft. In November 1975, DARPA awarded contracts to these companies to design and test models of low observable demonstrator aircraft. Early on, the U.S. Air Force assumed leadership of the effort. Following a competitive evaluation of large-scale radar cross section (RCS) models, which were used to validate the predicted very low radar signatures, the Air Force issued a contract in April 1976 to Lockheed Advanced Development Projects (ADP, also known as the "Skunk Works"). ADP was requested to produce and flight test

two low RCS technology demonstrator aircraft under a highly classified special access program known as "Have Blue." By mid-1979, the Have Blue aircraft had validated the concept of Lockheed's low RCS design approach by proving that its unconventional, faceted configuration could achieve acceptable flying characteristics as well as very low radar and infrared signatures in flight.

Before completing Have Blue's flight test program (but after flight performance and preliminary in-flight RCS testing had been accomplished), the Air Force, with strong support from the Department of Defense and key Congressional committees, initiated full-scale development of the F-117A under the Senior Trend program in November 1978. This highly concurrent and streamlined program applied the new low observables technologies and fielded a weapon system capable of highly survivable precision attacks against vital elements of an enemy's military, political, or economic assets. First flight was in June 1981, a limited F-117A initial operational capability was achieved by October 1983, and the aircraft subsequently played a prominent role in the air campaign against Iraq during Operation Desert Storm in early 1991.

The development of the F-117A was characterized by clear vision, strong leadership, highly responsive decision making, strong teamwork between the contractor and the Air Force, and a consistent focus on achieving an operational capability. These aspects of the effort are emphasized in this book along with potential application of the early Have Blue/Senior Trend experience to today's acquisition environment.

An overall timeline of the Have Blue technology demonstration and Senior Trend acquisition programs is shown in Fig. 1. A more detailed time table is presented in Appendix H.

BACKGROUND

Radar-directed weapons first became a serious threat to military aircraft during World War II. Correspondingly, the development of countermeasures against an enemy's use of radar became an important endeavor. Chaff and other countermeasures were developed, and the first attempts were made to reduce the radar signatures of aircraft. These early attempts, during the war and shortly afterward, consisted primarily of applying radar absorbent material (RAM) to all or part of a vehicle's surface. Specific examples of early radar signature reduction efforts are described in Appendix A. On the whole, the various "parasite treatments" (RAM, paints, and other add-ons to existing aircraft) did not produce any tactically significant reductions in radar detection range.

RADAR RANGE EQUATION

The primary reason for the lack of early success arose from the basic physics of the problem. The maximum range R_{max} at which a target can be

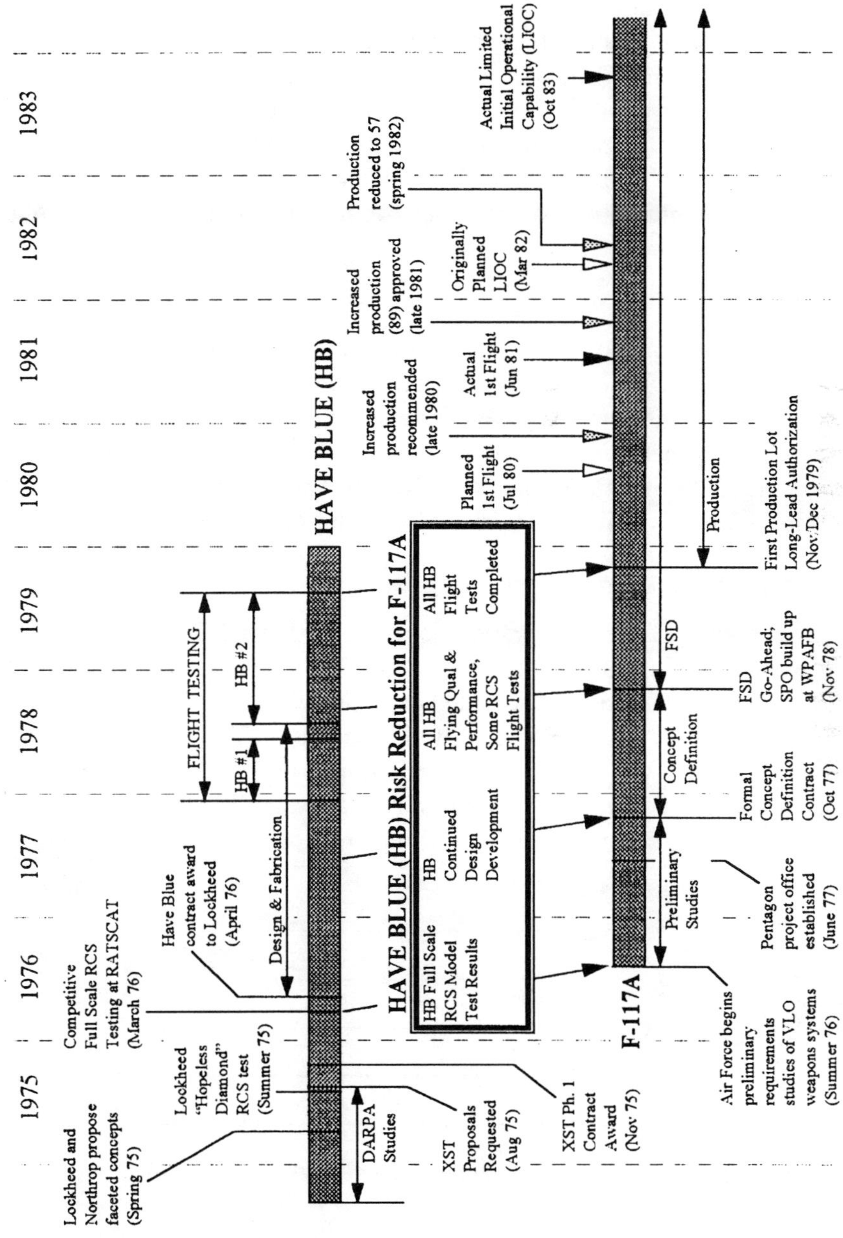

Fig. 1 Timeline, Have Blue concept studies through F-117A limited initial operational capability.

detected by radar is given by the well-known radar range equation

$$R_{max}=[(P_t/P_{min})(G^2\lambda^2\sigma/(4\pi)^3]^{\frac{1}{4}}$$

where P_t is the transmitted power, P_{min} is the minimum detectable returned signal power, G is the radar antenna gain, l is the wavelength, and s is the radar cross section (RCS) of the target (in square meters). The concept of radar cross section is explained in Appendix B. Because all objects of interest, from birds to geographical features, have RCS values that range within six or seven orders of magnitude, RCS is often expressed in decibels. In this form, the common unit of measurement is dBsm, or decibels relative to a one square meter reference RCS:

$$\sigma_{dBsm}=10 \cdot \log_{10}(\sigma)$$

The radar range equation shows that detection range is proportional to the fourth root of the radar cross section. Therefore, an extremely large change in RCS is needed to achieve any appreciable reduction in detection range. For example, to reduce the detection range by a factor of 10, it is necessary to reduce the target aircraft's RCS by a factor of 10^4, or 10,000. The dependence of detection range on RCS for a given hypothetical radar system is illustrated in Fig. 2.

LESSONS LEARNED FROM EARLY EXPERIENCE

The implication of the radar range equation is that RCS reduction is an extremely challenging endeavor. Thus, it is easy to see why the earliest efforts were not very successful. By the late 1950s, those working in the field realized that the very large RCS reductions necessary to achieve any operational benefit would not be accomplished simply by coating an otherwise conventional aircraft with RAM. Many physically small features of an aircraft generate radar returns that are still quite detectable. Not all such details

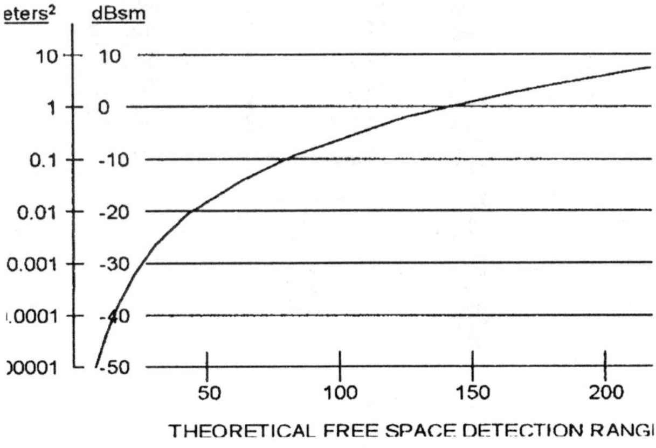

Fig. 2 Radar detection range as a function of RCS.

can be covered with RAM because this would interfere with their primary function. The most obvious examples are cockpit canopies and engine inlets. Additionally, locating a radar antenna behind a totally absorptive radome would be clearly unacceptable. Special design approaches and treatments are, therefore, necessary for any device that must pass matter or energy (including information) to or from an aircraft, to allow the device to perform its intended function while minimizing or eliminating the radar return that it generates.

Another consideration is that RAM is not 100% absorptive. Whenever electromagnetic waves encounter an obstacle, some radiation will be scattered. Therefore, an aircraft must also be shaped so as to direct the scattered radiation away from its source. A very low RCS must be "designed in" to an aircraft from the outset, with rigorous application of all three elements: RAM, special detail treatments, and overall shaping of the airframe and its components. From the late 1950s onward, efforts were made to incorporate these elements into various new aircraft designs. Some of the designs existed only as conceptual studies, while others were aircraft that were actually built.

During this time, basic research was also actively pursued. Under Army, Air Force, and Navy sponsorship, the major aircraft companies and other defense contractors, commercial and government laboratories, and several universities (including the Massachusetts Institute of Technology, the University of Michigan, and Ohio State University) all conducted research on various aspects of RCS reduction.

State of the Art Prior to Have Blue

The level of understanding that was achieved by the early 1970s in radar absorbent materials, special detail treatments, and airframe shaping for low RCS is discussed in the following sections.

Materials

Because most of the very early attempts at RCS reduction had focused on the use of RAM, this was perhaps the best understood area. By the early 1970s, a variety of materials had been developed and characterized. Specific purposes had been identified, such as reducing specular reflections (reflections normal to the surface), attenuating the waves that travel along a surface, or reducing edge returns. Specific approaches had been determined for each purpose. For specular RAM, reasonably good tools existed for designers to use to optimize a multilayer arrangement of absorbent materials:

> Computer programs have been developed for determining reflection, absorption and transmission of complex materials arrangements as a function of frequency, polarization and angle of incidence. These programs handle any number of layers of materials, each with arbitrary thickness

> and electrical properties. To aid design, a computer optimization program (AESOP) was combined with the electrical program....The combined program will vary each of the many parameters of a composite material until all design goals are met, if at all possible.[26]

Specular RAM was also in common use for commercial, nonaircraft applications, such as antenna pattern measurement facilities, used in designing and developing radio antennas. For these applications, materials were available with excellent absorption properties over a large bandwidth (i.e., a broad range of frequencies), and were even lightweight and low cost. However, they achieved their performance primarily through thicknesses up to several feet, making them clearly unacceptable for use on the surface of an aircraft. The chief challenges for aircraft RAM, then, were achieving good absorption performance over all required threat radar frequencies within acceptable thickness and weight constraints and developing materials and application methods that could withstand the severe mechanical, acoustic, and thermal environments encountered on aircraft.[26]

Additionally, complicated multilayer and/or tapered arrangements of absorbers could only be designed if the electrical properties of the basic ingredients were correctly understood. Thus, material characterization and quality control became important. Work progressed toward characterizing newly developed materials and also toward developing improved test techniques for characterizing new and existing materials. Furthermore, several anomalies between prediction and experiment had been traced to quality control problems; attention to manufacturing processes was necessary to ensure that test samples and actual production materials met their design specifications.[26]

Although materials for attenuating surface waves had not yet been as thoroughly studied, it was observed that specular RAM was fairly effective, if not optimal, for the purpose of attenuating surface waves.[26] With edge treatments, government laboratories and several aircraft companies (including Lockheed, North American, Northrop, Teledyne Ryan, and others) developed ways to construct edges for reduced radar return.[26, 105, 133, 186] Lockheed acquired extensive practical experience not only with the design and construction of such edges, but with operating and maintaining the edges on a high performance aircraft: the A-12 high altitude, Mach 3 strategic reconnaissance aircraft operated by the Central Intelligence Agency (CIA) in the 1960s. (This was the predecessor to the better-known SR-71 Blackbird.)

Detail Treatments

Inlets, exhausts, cockpits, antenna installations, propellers, rotors, and external stores were all recognized as major and sometimes dominant contributors to an aircraft's RCS.[26] Because of the significant challenges, much of the early signature reduction work was directed at applications for which one or more of these problems could be bypassed. For example, missiles do not require a cockpit, and if rocket-powered, do not require an inlet. However, effort was also directed at the problems of signature reduction on manned aircraft, and effective treatments were developed for many of the details already noted.

Very thin conductive coatings on a cockpit canopy were found to prevent backscatter from objects inside the cockpit (such as the pilot's helmet and the ejection seat). Various innovative concepts were developed for antennas and radomes. The University of Ohio Radiation Laboratory was one of the leaders in this area.[26, 123] Other research centers worked specifically on problems associated with propellers and helicopter rotors. For weapons, it was generally accepted that these would have to be carried internally.

Inlets and exhausts were perhaps the most challenging. Two main treatment approaches were identified for engine inlets: screens, or RAM lining on the inside of the inlet duct. Inlet screens were fitted to a Boeing B-47 bomber in 1960 and reduced its frontal RCS to a small fraction of the original value. However, there was a significant penalty in engine performance caused by the pressure loss of the air flowing through the screen. Experimental programs in the late 1960s achieved some success in developing inlet designs that achieved the same RCS benefits as screens but without the large aero/propulsive losses. RAM lining on inlet ducts also dates back to the early 1960s. In 1962 the entire fleet of North American Hound Dog air-to-surface missiles were retrofitted with RAM on the inlet spike and duct. Measurements indicated a substantial reduction in frontal sector RCS.[26]

Exhaust systems are conceptually similar to inlet ducts, but the problem is complicated by higher temperatures and airflow velocities. In general, potentially successful treatment of exhaust systems was only accomplished for fairly low-performance aircraft designs. In a 1972–1973 Quiet Attack aircraft study for the Office of Naval Research, McDonnell Douglas developed a special plug nozzle to conceal the aft face of the high-bypass, tip-driven turbofan engine.[257] A Teledyne Ryan "Mini-Remotely Piloted Vehicle" (mini-RPV), also designed in the early 1970s, concealed its ducted propeller with screens at the intake and exit of the duct.[256] As a partial solution to the exhaust problem, several aircraft designs had the fuselage and/or the tail configured to conceal the exhaust system from the most critical detection aspects.

Shaping

As of the early 1970s, shaping was the least-understood RCS reduction technique. Specular reflection (direct reflection from a surface that is perpendicular to the line of sight between the radar system and the target) had long been recognized as the strongest scattering mechanism. A 1947 textbook *Radar System Engineering* described various corner reflector configurations for situations when it is desirable to *increase* a target's RCS. Two of these, which use multiple reflections to generate a powerful specular return over a large range of viewing aspects, are illustrated in Fig. 3.[198]

The text also noted that

> If the problem is, on the contrary, to construct a target with a low radar cross section for most directions of approach...surfaces perpendicular to the probable directions of observation must be eliminated. Partial camouflage of targets has been achieved by proper shaping.[198]

Furthermore, the existence of other types of scattering was noted:

> Contributions to the cross section can be separated roughly into two parts. The first arises from flat or gently curved surfaces of considerable area (in square wavelengths) which are normal to the line joining the target and transmitter. Radiation returned from such surfaces may be properly classed as specular reflection; cross sections arising therefrom will be referred to as "specular cross sections." If any dimension of the target is smaller than a wavelength, or if the backward radiation arises from the secondary maxima in the diffraction pattern, the cross section will be referred to as a "diffraction cross section." When both types of backward scattering are present, the specular cross section is usually much greater.[198]

For nearly 25 years, this is about all that aircraft designers understood about shaping for low RCS. Accordingly, most early attempts to design low RCS aircraft, when shaping was considered at all, concentrated on eliminating surfaces that generated a specular return at the most likely detection aspects. The designs were characterized by slanted or chined fuselage sides and slanted vertical tails. Some examples are illustrated in Fig. 4.

These efforts and other related projects (described in Appendix A) achieved moderately low signatures. However, the cross sections resulting from the

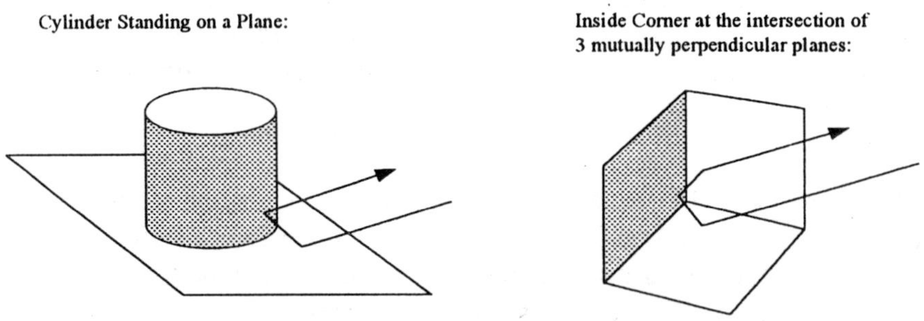

Fig. 3 Two types of corner reflector configurations.

remaining nonspecular sources were still high enough to be detected at a sufficient range for enemy defensive systems to react effectively. Further improvements depended on understanding and controlling the nonspecular sources. Theories that could predict the nonspecular scattering (referred to in the preceding quotation as the "diffraction cross section") had not progressed to usefulness for aircraft designers.* The eventual development of RCS prediction theory into a useful set of aircraft design tools is described in Appendix B.

Some attempts were made to develop shapes with very low radar cross sections through experimentation. Following a series of unsuccessful attempts to reduce the RCS of the U-2 aircraft, the Lockheed Skunk Works worked to determine what kinds of shapes should be used in a new aircraft design to achieve a low RCS. In 1975, Kelly Johnson (founder and, at that time, recently retired president and general manager of the Skunk Works) recalled some of these efforts:

> In order to evaluate the effect of overall shape on radar cross section, tests were run over a 10-year period of many different aircraft designs.... Elementary shapes were evaluated, such as disks of different contours, streamline shapes, and various bodies designed for producing low radar sections without consideration of their aerodynamic feasibility as aircraft. Fundamentally, a shape similar to flying saucers with a sharp edge and no protuberances has a very low radar cross section without any anti-radar treatment at all.... Attempts were made simultaneously to develop power plant installations and control systems which would make some of the low radar cross section objects fly satisfactorily.[105]

These efforts were apparently not pursued very far, however, prior to the Have Blue program. The A-12 and SR-71 configurations were derived primarily

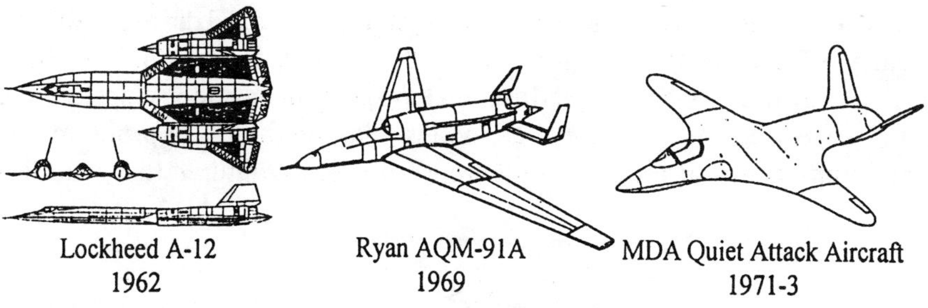

Fig. 4 Examples of pre-1975 reduced-RCS configurations.

*"Diffraction," as used here, refers to a variety of nonspecular scattering mechanisms. Diffraction also refers to a specific mechanism, namely the scattering from an edge whose cross section is either sharp or has a very small radius of curvature relative to the incident wavelength.

from aerodynamic considerations, incorporating RCS-reducing features only when there was no significant adverse impact on aerodynamics.

Teledyne Ryan also conducted experiments for developing low RCS aircraft shapes. The AQM-91A (one of the aircraft illustrated in Fig. 4) had been Teledyne Ryan's first attempt to design a low RCS aircraft from the ground up and had used the typical approach of orienting surface normals away from the critical aspects, together with RAM treatment of certain components. Subsequent experimentation led to abandoning a conventional wing–body–tail design and adopting a simpler, delta-wing concept as shown in Fig. 5.

Although this design and similar ones still did not meet certain RCS goals at all frequencies, they did provide the first credible indication that an aircraft's signatures could be reduced to the extent that some threat systems would not be able to detect or track the aircraft.[133, 234, 256] For this reason, the Teledyne Ryan accomplishments provided important impetus to further efforts.[188, 234]

INCREASING MILITARY INTEREST

Meanwhile, the subject of aircraft detection and tracking by radar became increasingly urgent. During the Vietnam War, radar-guided surface-to-air missiles and anti-aircraft guns, supplied by the Soviet Union, seriously restricted the ability of U.S. aircraft to perform their missions. By the later stages of the war, fewer than half of the aircraft involved in major U.S. air strikes carried weapons intended for primary targets. The rest included tankers, fighter escort, and, increasingly, aircraft dedicated to suppression of enemy air defenses. This included nonlethal (jamming) and lethal (anti-radiation missile) forms of suppression. Strike aircraft began to be equipped with on-board chaff dispensers and increasingly sophisticated jamming systems. Electronic countermeasure (ECM) techniques advanced rapidly during the war, but the state of the art in radar-directed threats also improved. Not only the capabilities of individual systems but also the variety of systems increased dramatically from the early SA-2 through the SA-3, -4, -5, -8, -10, -12, and so on. The various systems encompassed a range of different engagement envelopes (speed/range/altitude), frequencies, and guidance modes, making the ECM problem much more complex. An aircraft with inherently lower signatures that would not

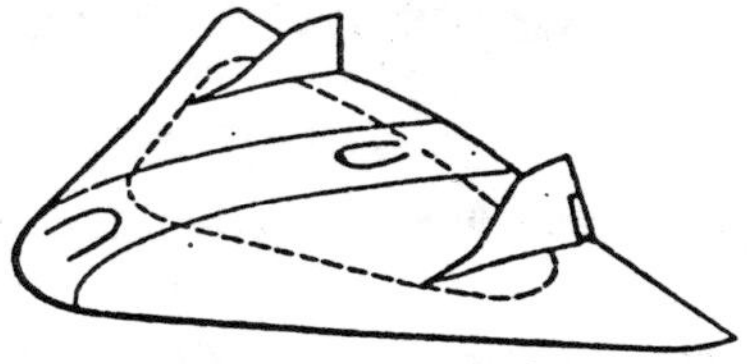

Fig. 5 Early 1970s Teledyne Ryan "low RCS vehicle" concept.[256]

have to jam or deceive the growing variety of potential threats would be a very appealing solution if it could be developed.

Accordingly, in 1971, a new Air Force prototyping study team recommended the development of a very low radar cross section test vehicle as part of a Department of Defense (DoD) -wide move toward advanced prototyping initiated by Deputy Secretary of Defense David Packard. The very low RCS test vehicle was proposed by Teledyne Ryan, and was one of six programs that the study team recommended from of a field of 45 candidates, noting that

> ...the capability to achieve extremely low radar cross sections has been demonstrated by large scale, nonflying test models....The capability to remain "invisible" to radar so radically changes the posture of most offensive and defensive systems that prototype testing to confirm this capability is warranted.[246]

Teledyne Ryan proposed to build and flight test three unmanned aircraft. The proposed design was almost certainly similar to the delta-wing shape, illustrated in Fig. 5. Gross weight would be approximately 3000 lb. The aircraft would be subsonic but capable of very high-altitude flight. Launch would be accomplished from a modified C-130 with recovery by the existing mid-air retrieval system (MARS) as used by their earlier AQM-91A Compass Arrow and other drones.[246]

The other projects recommended by the study team were an advanced medium STOL transport (AMST), a large tanker aircraft, a Lightweight Fighter (LWF), a Quiet Aircraft, and an Advanced RPV. The AMST and LWF programs were approved and ultimately led to the flight testing of the Boeing YC-14 and McDonnell Douglas YC-15 STOL demonstrators and the General Dynamics YF-16 and Northrop YF-17 lightweight fighters respectively.[13, 131] The very low RCS vehicle was not pursued at that time.

A further demonstration of the lethality of radar-guided air defense systems occurred in October 1973. In the Yom Kippur War, Israel lost more than 100 combat aircraft—a substantial fraction of its front line fighting strength—in just 18 days, most of them to Soviet-built radar-guided surface-to-air missiles and guns operated by Egypt and Syria. This was particularly disconcerting because Israel was using up-to-date Western aircraft, radar countermeasures, and tactics. The complementary elements of the Soviet Integrated Air Defense System (long-range systems with large, fixed radars, coupled with shorter range mobile missile and gun systems) rendered them not only extremely lethal but also nearly invulnerable to attack from the air, and they were only overcome by unusual and inefficient methods. On the northern front, Syria simply ran out of missiles following two days of intensive Israeli air action that included many "fake" attacks intended specifically to draw fire. In the south, the Egyptian air defense sites were not neutralized until fairly late in the war, and then not from the air but by Israeli ground forces.[72] The experience of this war led to serious concerns. Predictions were that the U.S. Air Force would be decimated in about two and a half weeks if there were a full-scale war against the Soviet Union in Central Europe.[115]

With aircraft survivability rapidly becoming an extremely critical issue, the Air Force Avionics Laboratory (AFAL) and the Air Force Materials Laboratory (AFML), both of which were part of the Air Force Wright Aeronautical Laboratories, jointly sponsored the 1975 Radar Camouflage Symposium. The symposium was held in October 1975 at Wright–Patterson Air Force Base near Dayton, Ohio. C. E. (Chuck) Myers, Jr., assistant director for air warfare, Office of the Director of Defense Research and Engineering, gave the keynote address. This was followed by an overview of recent progress in radar camouflage, presented by William F. Bahret, acting chief of the passive ECM Branch within the Electronic Warfare Division of AFAL, who was an active proponent of early low observables research and development efforts, having been involved in radar camouflage work since at least 1960.[26, 118, 204] Overall, 23 papers were presented to an audience of more than 250 attendees from industry, academia, and government. Topics included the benefits of radar camouflage; various types of radar absorbent materials; design approaches; and applications to aircraft, missiles, and drones.

One purpose of the meeting was to generate an increased awareness of progress to date and to stimulate further activity in the field of RCS reduction (or "radar camouflage"). The general theme was that substantial reductions in RCS could be achieved "by simply being concerned about the matter from the outset of design," and that such reductions, in conjunction with ECM, would offer improved survivability with little or no penalty, and possibly even a cost savings accounting for the fact that a reduced-RCS aircraft can use smaller, lighter, and less powerful ECM systems.[26] However, the underlying assumption in most of the papers presented was that RCS reduction would complement, not replace, ECM. The Have Blue program, for which the preliminary stages had already been underway for about a year, was founded upon a completely different approach.*

---

*Several key participants in the Experimental Survivable Testbed program (the early name for what became Have Blue) were present, including Kelly Johnson, Mel George, and Ed Lovick from Lockheed; John Cashen and Stanley Locus from Northrop; Bill Elsner from ASD; and Ken Perko from DARPA. Johnson presented a paper on "Reduction of Radar Cross Section of Large High Altitude Aircraft," in which he described Lockheed's experiences with the U-2, A-12, and SR-71. The XST program was not explicitly mentioned in the conference presentations. William Bahret noted in his introductory paper: "The reader is cautioned that the examples to be given by no means represent all applications [of radar camouflage] to date. Indeed, a number of the most important applications are too sensitive for inclusion here." Bahret concluded by noting that the examples he had already discussed, "as well as several aircraft of equal novelty currently on drawing boards, are irrefutable proof" of the benefit that can be achieved by designing a system from the outset for minimum RCS; and that "demonstration programs are in progress today for the express purpose of removing any remaining doubts of the practicality and feasibility [of radar camouflage]."[26]

Chapter 2

Have Blue

DARPA Study Program, 1974–1975

Objectives

During 1974, the Defense Advanced Research Projects Agency (DARPA) initiated, with U.S. Air Force participation, a program to study and possibly demonstrate the concept of a very low observable military aircraft. DARPA was founded in the 1950s to support high risk and potentially high payoff technology with multiservice applications. The DARPA studies had two basic objectives: designed to identify signature levels that would permit a tactical aircraft to avoid detection (primary emphasis on radar, also infrared, with visual and acoustic detection as tertiary considerations only) and to define a technical approach for achieving such levels.[151, 188, 204]

The study was initiated and led by Ken Perko, who had recently come to DARPA from the Remotely Piloted Vehicle System Program Office (RPV SPO) in the Air Force Systems Command's Aeronautical Systems Division (ASD), Wright–Patterson Air Force Base, Ohio.[188, 234] Perko was recruited to DARPA in 1974 by Kent Kresa,* then head of DARPA's Tactical Technology Office (TTO). Perko's job was to build up a tactical air division within the

*Kresa is currently chairman and chief executive officer of Northrop Grumman Corporation.

TTO at DARPA. Pentagon Summer Studies that year had focused on Central European scenarios, in which it was projected that NATO forces would be defeated if there were a full-scale war against the Soviet Bloc countries. The capabilities of the Soviet Integrated Air Defense Systems (IADS) were a major factor in this outcome. It was clear that something had to be done to improve the survivability of tactical aircraft. Further developments in electronic countermeasures (ECM) were considered by some to be the answer. Signature reduction, by itself, was not yet known to be a viable solution.[188]

As noted in Chapter 1, improving an aircraft's survivability by reducing its signature is a very demanding problem. The basic physics of the problem are such that to reduce detection range by a factor of 10, it is necessary to reduce the target vehicle's radar cross section (RCS) by a factor of 10,000. Such low signatures seemed completely out of technological reach. For this reason, many projects focused on signature reduction only as a complement to ECM. However, ECM has the drawback that it does not prevent detection and may actually alert the enemy to the fact that something is approaching. Perko decided to see whether it was possible to reduce signatures sufficiently to avoid detection altogether.[188, 204] Tests conducted by the RPV SPO, in conjunction with Teledyne Ryan and other companies, were at that time beginning to indicate that this might be possible. There appeared to be certain thresholds of target RCS below which an airborne target would become extremely difficult to detect and track. Perko was aware of these promising developments as a result of his prior work in the RPV SPO.[188, 234]

Perko requested White Papers from five companies—Northrop, McDonnell Douglas, General Dynamics, Fairchild, and Grumman—addressing two topics. First, what were the signature thresholds that an air vehicle would have to achieve to be essentially undetectable at an operationally useful range? Second, what were the capabilities of each company to design and build an aircraft with the necessary low signatures? Fairchild and Grumman did not express any interest in participating. The General Dynamics response emphasized countermeasures and had little substantive technical content regarding signature reduction. McDonnell Douglas and Northrop responded indicating a good understanding of the problem and at least some technical capability for developing a reduced-signature air vehicle. In late 1974 these two companies were awarded contracts of approximately $100,000 each to conduct further studies. Radar experts at Hughes Aircraft Company were also asked to provide input. These initial studies were classified Confidential.[188, *] Contracting was conducted through the RPV SPO, which also provided technical assistance. Bill Elsner was the primary U.S. Air Force technical expert on the program at this stage.[188, 234]

*There are three major levels of security classification: Confidential, Secret, and Top Secret. Confidential is the lowest of the three.

McDonnell Douglas was the first to identify what appeared to be the appropriate RCS thresholds. Shortly afterward, these were confirmed by Hughes. In the spring of 1975, DARPA established these values as goals for the program, and challenged the study participants to find ways to achieve them.[188, 197, 234]

Lockheed Involvement

Lockheed had not been invited to participate in the studies, simply because the company was not considered to be active in tactical aircraft development at that time. Lockheed had not built a new fighter-type aircraft in more than 10 years. In fact, all aircraft development activity was slow at Lockheed at this time, and so the company was very interested in any prospect that might lead to new business. Ed Martin, a veteran of the Lockheed Skunk Works who at that time was science and engineering director for the entire Lockheed California Company (of which the Skunk Works was a part), probed his contacts at the Pentagon and at Wright–Patterson Air Force Base for any promising activities and came across the DARPA project almost by accident in late 1974 or early 1975.[204]

To provide evidence that Lockheed had experience in low observables and should, therefore, be involved in the project, Kelly Johnson* obtained permission from the CIA to brief DARPA on previous, highly classified work performed by Lockheed Advanced Development Projects (Skunk Works) for the CIA with the A-12 and D-21 reconnaissance aircraft.[151, 175, 234] Ed Martin and Ben Rich made the presentations to DARPA, and as a result, "Lockheed received permission from the government...to provide an input to the studies program consistent with the schedule requirements of the other participants."[234] Lockheed was also given access to technical reports and other information relevant to the studies, which had already been provided to the other participants.[186]

Having secured a place for Lockheed in the competition, Martin recruited Dick Scherrer in February 1975 to head a preliminary design group at the Skunk Works that would put together Lockheed's entry in the competition. Scherrer was the Lockheed "white world"† Advanced Concepts Department manager and had been the proposal manager for an initial study of what became the Lockheed S-3A Viking antisubmarine aircraft.[151, 204] Scherrer made inquiries within the Skunk Works to identify a theoretical foundation on which he could base a low RCS design. However, Lockheed did not have any rigorous

---

*Kelly Johnson was the director of LADP from its formation in 1943 until his retirement in January 1975. At that time, Ben Rich became the director of the Skunk Works, while Johnson stayed on as a consultant.

†"White world," in this context, refers to programs whose existence is openly acknowledged. Conversely, all aspects of "black world" programs, including their existence, are totally covert.

theory or predictive techniques that could be used to design an aircraft from the beginning for low RCS.

The Skunk Works had experience in RCS reduction with the A-12 reconnaissance aircraft and the subsequent D-21 drone using a combination of shaping, special materials and construction techniques, and extensive experimental work to identify and correct the major scattering sources. However, the shaping was largely intuitive or at best was based on some empirical rules to establish such parameters as minimum surface slopes (for example, the slant of the vertical fins on the A-12). Furthermore, RCS considerations were never allowed to compromise the aerodynamics or propulsion performance of these high-altitude, Mach 3+ designs.[34] Nevertheless, Scherrer was told by the Skunk Works' RCS group to "use the D-21 as a design guide."[204]

Scherrer received one slightly more promising lead through Warren Gilmour, an acquaintance of Scherrer's who conducted operational analyses at the Skunk Works. Gilmour had discussed the subject of RCS theory with Denys Overholser, head of the Skunk Works' computational group and a former radome designer. When presented with the question of how to design an aircraft for minimum RCS, Overholser referred to an idea that had come out of discussions years earlier between Overholser and his retired former boss, Bill Schroeder. Schroeder, a mathematician, had been "Kelly Johnson's expert for any kind of analytical problem," the forerunner of Overholser's computational team.[186] Schroeder had trained Overholser, and during that time the two had discussed the mathematics and physics of optical scattering. They had come to the conclusion that detectable signatures could be minimized with a shape that was made up of the smallest possible number of properly oriented flat panels. Furthermore, Schroeder thought that he could find an explicit mathematical expression for the reflection from a triangular flat panel, thereby permitting the RCS of such shapes to be calculated analytically.[186, 204]

Scherrer had been involved many years earlier at NACA with wind tunnel tests of wings that were made up of flat panels. Configurations such as those shown in Fig. 6 were studied extensively by NACA and industry during the late 1940s and early 1950s.[193, 204, 249] Double-wedge airfoils were used on these configurations, rather than conventional curved airfoils, because the double-

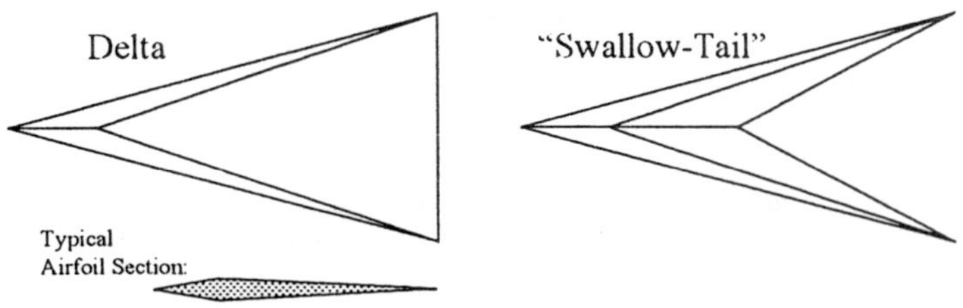

Fig. 6 Typical supersonic faceted wings, late 1940s.[193, 249]

wedge sections were known (from aerodynamic theory) to have certain advantages at supersonic speeds. It was also believed that the flight characteristics at subsonic speeds, although not optimal, would at least be acceptable.

Scherrer drew up a preliminary low RCS shape based on the faceted delta wings of his earlier experience together with input from Overholser on what the surface slopes and edge orientations should be to minimize RCS. Although this was only a preliminary shape, some thickness was provided so that engines, crew accommodation, and other necessary items could eventually be incorporated. The initial configuration resembled a flattened double pyramid, nearly flat on the bottom, and without any distinct wings or tail surfaces, as shown in Fig. 7.

Overholser contacted Schroeder to discuss completing the solution to the RCS equations for a plate so that Overholser's group could eventually program the results. By April, Schroeder had been hired as a consultant to accomplish this task. Kenneth Watson, who had worked with Scherrer in the Advanced Concepts Department, was also brought in as senior lead airplane designer. According to Scherrer,[204] "It was Kenny's task to make an airplane, with all its internals, stuffed into the external shape being evolved by [myself] and Denys—a frightening challenge to any experienced airplane designer."

Warren Gilmour conducted the requirements analysis to determine what signature levels would have to be achieved to satisfy the DARPA objectives and win the competition. Gilmour simulated some one-on-one encounters between aircraft and radar-guided missile and gun systems, and it was obvious from the results of these simulations, together with continuing guidance from DARPA, that D-21 RCS levels would not be adequate.[186, 188, 204] This led Scherrer to follow the less-accepted faceting approach against the prevailing "wisdom" in the Skunk Works.

"I still view my role in Have Blue with some confusion," recalls Scherrer. "Although my assignment from Martin was to win [the DARPA competition], that responsibility was never backed up with any assignment of authority. I could not assign a task to anyone in the Skunk Works. I had to convince each person that

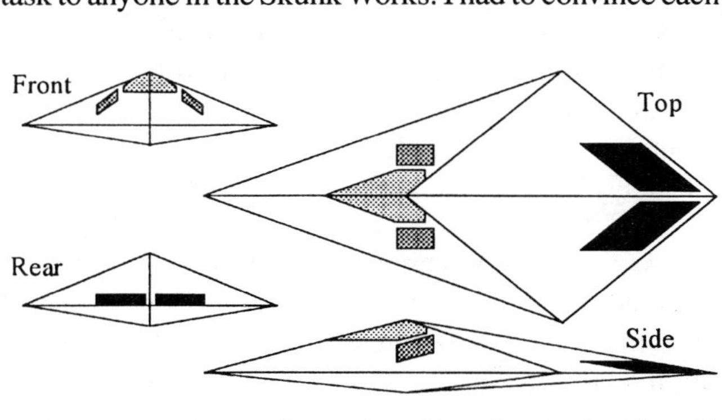

Fig. 7 Initial pre-Have Blue configuration. (Based on Refs. 186, 188, and 203.)

the task was worthwhile, in their terms, which in most cases led to improved task statements." Dick Cantrell, the Skunk Works' senior aerodynamicist, who was sharing his office with Scherrer, accepted responsibility for the aerodynamics and flight control effort, which would ensure that the new design could be made to fly.[197, 204]

Two activities began to take place in parallel. As Schroeder's mathematical results became available, Overholser and his group began to write a computer program that could evaluate the RCS of candidate designs to aid the preliminary designers in refining the configuration. Concurrently, because they could not wait for the program to be developed, Scherrer, Watson, and Overholser proceeded to refine the design as well as they could with the tools at hand. Ed Lovick, who had been involved with earlier RCS work on the A-12 and SR-71, provided Scherrer with sets of two-dimensional reflection plots that included "side lobes" (reflection strength at nonspecular angles). These plots, together with "astute guidance" from Dick Cantrell and his flight sciences group, helped the designers to develop a shape to satisfy both signature and flight requirements.[186, 204]

The urgency was a result of the tight DARPA competition schedule. In the spring of 1975 Lockheed, Northrop, and McDonnell Douglas were the three participants in the program of which Lockheed was the newcomer. A downselect was planned for late summer or early fall of 1975, and Lockheed wanted to be sure that it was one of the winners.[186, 204]

For this reason, Overholser and his team were given approximately six weeks to write their RCS prediction program.[186, 197, 204] Schroeder worked on the basic equations forming the heart of the program. Overholser double-checked Schroeder's work and wrote some of the computer code when he was not busy with the aircraft designers. Two other engineers working for Overholser handled the aircraft geometry input as well as plotting/output portions of the program. "ECHO 1," as the initial version of the program became known, was actually running within the six weeks originally specified, "due to Denys living with it night and day," recalls Scherrer.[204] However, it should be pointed out that ECHO 1 was a relatively simple RCS prediction program. According to Overholser, ECHO 1 represented a typical Skunk Works problem-solving strategy of reaching a "95% solution" as quickly as possible. In spite of its limitations, the program was sufficiently accurate to be a useful design tool. Consistent with the early 1970s state-of-the-art in computing, the input consisted of decks of punched data cards, making it a somewhat laborious process to input the geometry for each new design variation that had to be analyzed.[186] However, output was more convenient, consisting of polar plots showing the RCS as a function of aspect angle for each frequency and polarization analyzed.[204]

ECHO 1 Approach and Theoretical Foundation. Most methods of RCS calculation for bodies that are large relative to the incident radar wavelength follow a two-step approach. First, calculate the electrical currents induced on the surface of the target by the incident radiation and second, perform a surface integral, or summing of the effects of the currents over the entire surface to find the scattered radiation.* To simplify the determination of surface currents, ECHO 1 assumed that the vehicle's surface was metallic and therefore a perfect electrical conductor. ECHO 1 also used an approximate technique known as physical optics (PO) in which the current at every point on the surface is assumed to depend only on the surface slope at that point. Effects of neighboring portions of the surface are not considered. This greatly simplifies the first step—calculating the currents—but the second step, the integration, was not trivial. Schroeder's accomplishment was to find the explicit expression for the value of the integral for a triangular flat panel. The ECHO 1 computer program automated the calculation and summed the results for a complete configuration that was made of, or could be approximated by, a number of flat panels.[34, 186, 204]

Schroeder's solution provided information on the strength of the surface returns even for nonspecular angles when the surface in question is not perpendicular to the incident radiation. This told the designers how much each surface would have to be slanted to achieve a particular RCS at a given aspect. Earlier efforts at low RCS aircraft design had relied on intuition or at best empirical rules of thumb to select surface slopes. Now the designers had a quantitative tool. The ECHO 1 solution also included contributions from edges, because the edges form the boundaries of the surface integral. So the designers could use ECHO 1 to manage edge orientations. Most previous attempts to design low RCS aircraft had failed to do this. The ability, provided by ECHO 1, to calculate edge returns was a key step forward in low RCS aircraft design.[186, 204]

However, the edge contributions calculated by ECHO 1 were not exactly correct, because the electrical currents near the edge were still assumed to be the same as everywhere else on the surface. In reality, there is a region near any sharp edge where the electrical currents are affected by the presence of the edge. Intuitively, one can visualize that if the surface currents are interrupted by a sharp edge, there is no place left for the electrical charge to go except *along* the edge. The radiation produced by these currents along the edge is known as diffraction. Some time after the initial development of ECHO,

*There are, in addition, certain phenomena such as multiple reflection, traveling waves, and creeping waves, which are not calculated by this approach and therefore require additional calculations. A more complete discussion of electromagnetic scattering theory, and the development of RCS prediction techniques, is presented in Appendix B.

Denys Overholser came across *Method of Edge Waves in the Physical Theory of Diffraction* by P. I. Ufimtsev, originally published as an unclassified technical paper in the Soviet Union in 1962 and translated by the Air Force Systems Command's Foreign Technology Division in September 1971. Overholser recognized that this theory addressed an important shortcoming of ECHO 1. He was eventually able to incorporate diffraction calculations based on Ufimtsev's theory into later versions of the ECHO program. However, during the early design stages of XST/ Have Blue, this feature was not yet available.[34, 186, 204]

Nevertheless, ECHO 1 was surprisingly accurate, in part due to pure luck. This is because Lockheed used special edge construction techniques, developed during the A-12 program, on the major planform edges of the aircraft, which were fairly effective at reducing the electrical currents in the edges. The physical mechanism of diffraction, on the other hand, tends to increase the currents in the edges. Fortuitously, the two effects approximately canceled each other and so the original version of ECHO, which did not account for either effect, was still fairly accurate. Later, by the time the F-117A was designed, ECHO had been upgraded to calculate the effects of edge diffraction and to include, at least empirically, corrections for the effects of special materials and coatings.[34, 186]

RCS Model Tests. Early in the concurrent process of aircraft design and ECHO development, the small team that was involved requested funding for an RCS test in the Skunk Works' indoor anechoic radar chamber. However, there was considerable opposition among some of the established members of the Skunk Works, who began to refer to the new airplane as the "Hopeless Diamond." Many believed that the earlier D-21 represented the best low observable design that could practically be accomplished and resented the new approach because it did not look anything like an airplane, and also because it was being proposed by Scherrer, who was an "outsider" to the Skunk Works. The proposed test of the Hopeless Diamond might not have taken place except for the strong support of Ed Martin, who from his position as corporate science and engineering director had perhaps less emotional attachment to past Skunk Works achievements and a better perspective on what was good for the company in the long run.[204]

With Martin's assistance, $25,000 was provided for the construction and testing of a wooden shape coated with metal foil to provide a conductive surface. The first tests took place around June 1975.[186, 204] Test results at the Skunk Works' indoor range showed that all the RCS "spikes" were exactly as predicted by ECHO.[204] The nulls between the spikes were too low to be accurately measured. The model was then taken to McDonnell Douglas' Grey

Butte outdoor RCS range, which had improved capabilities for measuring low signatures, and the team was able to measure lower RCS values than before in the "null" regions between the spikes. These values were still higher than predicted and were believed to be caused by the conic Styrofoam column used for the model support, rather than by the model itself (which later proved to be true). However, to the extent of the accuracy of the data that was obtained, both the indoor and outdoor RCS tests agreed well with the ECHO predictions. This provided confidence in the computer program and in the faceted design concept.[186, 204]

Design Evolution. Once it was established that the faceted shape did indeed offer a way to achieve very low signatures, the design effort picked up momentum. Leo Celniker was brought in from the "white world" and became chief engineer.[34] As is typically the case, the preliminary design group became the focal point for inputs from the various engineering disciplines (aerodynamics, flight control, structures, etc.) along with the relatively new discipline of low observables. The most significant kinds of input at this stage were from the aerodynamics and flight control engineers. To improve the vehicle's lift-to-drag ratio, the region outboard of the engine inlets was thinned down, resulting in surfaces that could actually be identified as wings. This quickly saturated the number of facets that ECHO could handle, and so Overholser's group was constantly being pushed to increase the program's capability.[204] Eventually, the wings were extended outward, changing the planform from the original diamond shape to the notched-out delta that characterizes the Have Blue and subsequent F-117 aircraft. Tail surfaces were also added. The general progression from the Hopeless Diamond to a configuration that closely resembles Have Blue is shown in Fig. 8.

The design also was defined in progressively greater detail. Items such as the engine installation, which had not received a great deal of thought in the very early stages, began to be considered. By 1975, screens or grids were a known way to reduce radar backscatter from engine installations, having been tried in various laboratory and flight test experiments as early as 1960.[26] The theory is that radar waves will not pass through an array of holes whose size and spacing is small relative to the incident wavelength. Putting a screen or grid over the inlet can therefore "hide" the inlet duct and engine face. To the designers of the XST, this meant that they could treat the plane of the inlet grid as just another one of the facets on the vehicle's surface. However, screens are also a well-known way to produce a pressure loss in any fluid that flows through them, which could have a devastating effect on the installed engine thrust. Grids and other configurations had been tried, but generally did not achieve acceptable flow qualities over a useful range of flight conditions.[26, 34]

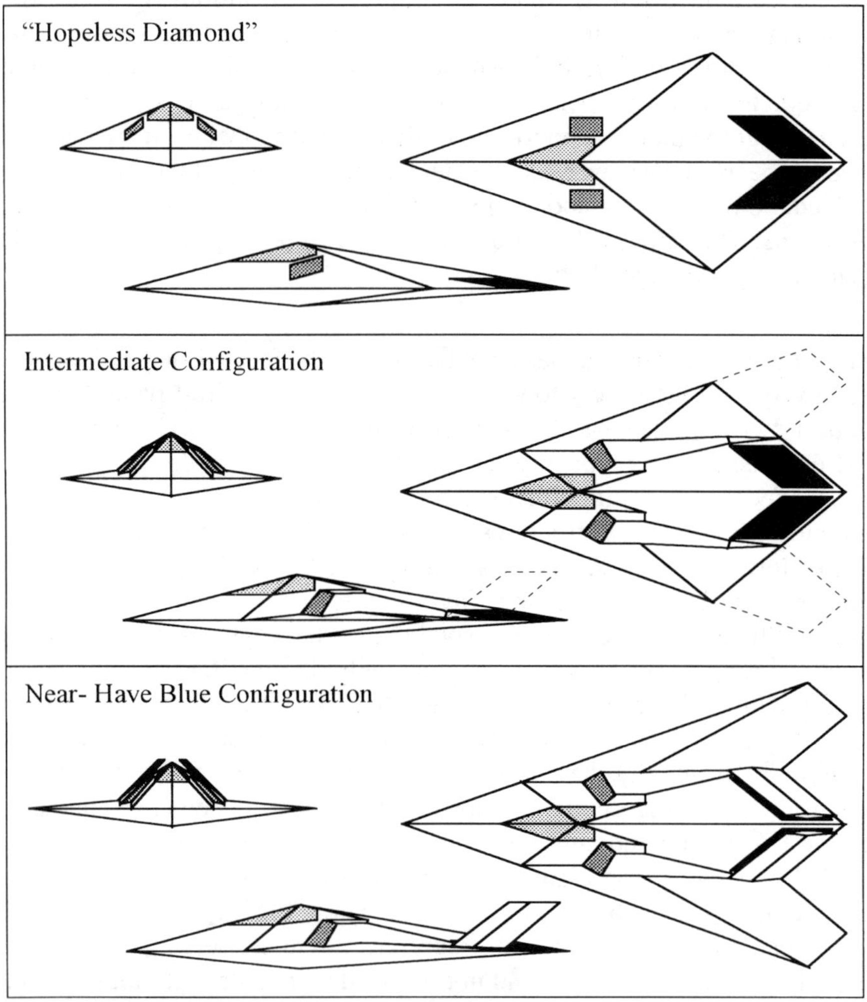

Fig. 8 Progression from Hopeless Diamond to early XST (notional, based on Refs. 175, 203, 186, and 204).

For this reason, Alan Brown was assigned to the project in June or July of 1975, from Lockheed California's main plant. Brown had previously headed a team that traveled to England to resolve some propulsion integration problems with the Rolls-Royce engines for the Lockheed L1011 commercial jet transport. He had also taught graduate-level fluid dynamics.[186] Brown's job in the XST project was to develop an inlet design that would have acceptable flow characteristics in addition to the desired electromagnetic properties. To gain an understanding of both sides of the problem, he sat down with Overholser, as most engineers on the project had done, for an explanation of

RCS principals. He picked this up very quickly, in part because of his prior experience. The governing equations of fluid dynamics have a lot in common with the governing equations of electromagnetic fields and waves. Brown, whom Overholser describes as "brilliant," could see at a glance what the mathematics meant. He not only helped solve the problem at hand, but quickly assumed the primary responsibility for low observables design integration, while Overholser handled the "basic science" aspects of RCS prediction and measurement technique. Brown went on to become chief engineer for Have Blue; chief engineer and program manager for the F-117A; and eventually head of Lockheed's research groups responsible for radar and infrared signatures, acoustics, and materials.[34, 186]

Proposals for what became known as the Experimental Survivable Testbed (XST) program were informally requested by DARPA in late summer of 1975, and were due in August or September.[186, 188] Signature goals "were derived from the 1974 –1975 studies and directed to the XST bidders as challenges for the program."[234]

At the last minute, Lockheed changed the sweep angle on their wing trailing edges. Large sweep angles on the trailing edges, and associated control surfaces, were desirable to keep the RCS spikes as far as possible from the nose-on aspect (remembering that any edge produces a return, both forward and aft, at the aspects that are perpendicular to the edge). However, highly swept control surface hinge lines result in diminished control effectiveness and increased cross-axis coupling (for example, yawing moments due to aileron deflection). The designers had initially selected what they felt was a reasonable compromise. However, as DARPA and the Air Force worked to define a way to evaluate the competing XST designs, they developed an algorithm for reducing all of the signature data to a single number. Signature values at different frequencies and aspects would be multiplied by various weighting factors, and then added together. The frontal sector, defined to extend 45 deg either side of nose-on, had the highest weighting factor. The designers, therefore, increased the trailing edge sweep to 48 deg so that the signature spike associated with the trailing edges would fall completely outside of the frontal sector. The impact of this decision on control effectiveness was accepted.[34, 204]

By the time proposals were due, substantial revisions to the Hopeless Diamond had been made, and the computer group had caught up with the designers in modeling the necessary number of facets. ECHO RCS predictions were made for the revised design. The proposal submitted by Lockheed included the predicted and measured signature data for the Hopeless Diamond, along with the predicted data for the new, "somewhat flyable" configuration.[186]

Thus, within three or four months of its inception, the Hopeless Diamond had become a viable preliminary aircraft design with an RCS that was potentially several orders of magnitude lower than any existing airplane. Achieving full potential meant complementing the aircraft's shape by applying appropriate radar absorbing materials (RAM) to the surfaces and edges and ensuring that details (such as air inlets, canopy, and access panels) were treated to behave electromagnetically as if they were a part of the vehicle's basic surface. Otherwise, the analytical predictions would be useless.

NORTHROP ENTRY

Northrop's entry in the XST competition is illustrated in Fig. 9. Like Lockheed's XST design, the Northrop entry consisted primarily (although not exclusively) of flat surfaces with straight edges.

Like Lockheed, Northrop had developed a capability to perform computational RCS prediction. However, Northrop's capability was achieved in a very different way. Northrop had been working on RCS prediction techniques at least since 1966, along with other aspects of RCS reduction, under the sponsorship of the Air Force Avionics Laboratory and other agencies, including the U.S. Army Missile Command.[39, 132, 176, 177, 185] Because the work was done on government contracts, the reports were made available to other defense contractors. For example, a Boeing RCS report of 1974[118] refers to the Northrop documents. Northrop's RCS prediction research group included S. Stanley Locus, Fred K. Oshiro, H. C. Heath, J. Randall Coleman, and Kenneth M. Mitzner. John Cashen headed the Northrop XST design team, which also included veteran aircraft designer Irv Waaland.[228]

In 1970 Northrop submitted user's guides and source code for a set of computer programs that calculated the RCS of various types of geometric shapes, including cavities (such as inlet ducts), with either perfectly conductive or RAM-coated surfaces.[185] For bodies with edges, P. I. Ufimtsev's now-well-known physical theory of diffraction (PTD) was used, along with various other applicable methods.[185] By 1974 the same research group had developed a "second-generation" computer program called GENSCAT, which

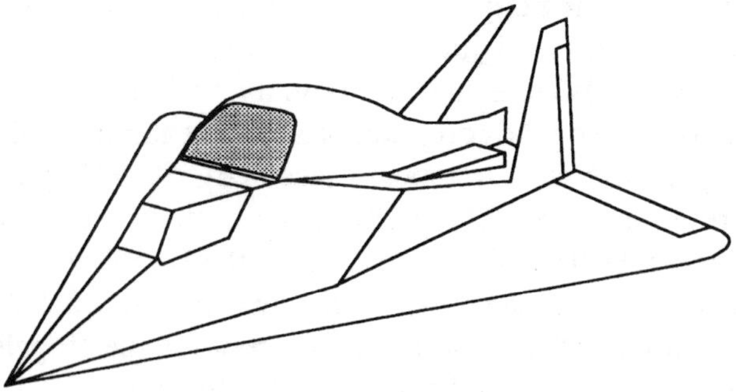

Fig. 9 Northrop's proposed XST.[104]

encompassed all of the earlier capabilities, together with several practical and theoretical enhancements, in a single program. Two particularly important theoretical contributions by Mitzner included an explicit mathematical expression for the physical optics scattering from a polygonal flat panel (similar to what Bill Schroeder accomplished at Lockheed) and a method called incremental length diffraction coefficients (ILDC) that extended the applicability of Ufimtsev's diffraction theory to arbitrary scattering aspects.[176, 177] RCS prediction theory is discussed further in Appendix B.

Thus, at the time the pre-XST studies began, Northrop actually had a much more comprehensive RCS prediction capability than Lockheed. As a test case, GENSCAT was used to calculate the RCS of an F-4C aircraft, and the results were fairly accurate. The time required to input the design geometry of the F-4C was substantial—three weeks to extract coordinate data from F-4 design drawings, and another seven days to keypunch the data and run programs that fit surfaces to the coordinates.[132] However, the XST configuration seemed simpler, and the turnaround time for analyzing variations of this design was probably manageable.

Nevertheless, it must be emphasized that GENSCAT, like ECHO, did not design airplanes. The computer program only calculated the RCS of an aircraft design that was given to it. Designing the airplane to begin with, and using GENSCAT results to improve the design, still required expert human designers who knew the underlying physics. In this regard, the understanding that was gained while developing GENSCAT may have been more valuable than the computer program itself.

Testing was also an important part of Northrop's effort. Material samples, specific components, and models of complete configurations were tested throughout the design process.[35] Interestingly, Northrop, like Lockheed, at times made use of McDonnell Douglas' Microwave Measurement Facility at Grey Butte, which was apparently one of the best facilities available at that time, for measuring very low RCS values.[39]

According to Ken Perko, Northrop proposed a faceted aircraft during the spring of 1975, approximately the same time as Lockheed. By the summer of 1975, Northrop had reliable indications that their design probably would achieve the RCS goals, which had been arrived at in the initial stages of the program.[188]

McDonnell Douglas Entry

According to Perko, McDonnell Douglas was the first of the study participants to arrive at a determination of what the RCS levels had to be.[188] McDonnell Douglas also had at their Grey Butte Microwave Measurement Facility one of the best test ranges in existence for measuring very low RCS values. As already noted, both Lockheed and Northrop conducted various tests at Grey Butte.[39, 126, 186] However, McDonnell Douglas was coming up short in the ability to design an aircraft that could achieve the desired RCS levels.[188]

A "Quiet Attack" aircraft, designed by McDonnell Douglas in 1973 on a contract with the Office of Naval Research (ONR), is illustrated in Fig. 10. The design was intended to avoid detection through reduced radar, infrared, acoustic and visual signatures. Infrared signature was reduced through the use of a high-bypass turbofan engine with a special plug nozzle to conceal the hot engine parts, which also reduced aft sector RCS. Acoustic signature was minimized by the high-bypass engine and a very slow flight speed ("quiet speed" of 112 kn, with maximum speed of 430 kn). So-called Yehudi Lights, or airframe-mounted lights that would match the ambient background lighting level were intended to reduce visual signature.[257, *]

The designers considered the following features to be essential for low RCS[257]: continuously curving leading and trailing edges on wing and tail surfaces, blending of the wing/fuselage junction, large leading and trailing edge fillets, vee-tails, engine inlets on upper wing surface (together with plug nozzle already mentioned), and RAM.

As is typical of pre-1975, low RCS design efforts, following these guidelines resulted in the elimination of the largest scattering sources, i.e., engine returns and specular reflection from the vehicle's surface. Using continuously curving planform edges, however, has some interesting results. Long straight edges do produce a spike in the radar signature at the aspects that are perpendicular to the edge. The longer the edge, the higher the spike. A curved edge does not produce such a large return, and this is doubtless the reason for the curved edges of the McDonnell Douglas aircraft. The return from a curved edge is still generally detectable, though, and the drawback is that there is a continuum of viewing aspects that are perpendicular to some point on the edge. The use of a single curve that wraps around the entire aircraft virtually guarantees a detectable edge return from almost any viewing aspect.

According to Ken Perko, McDonnell Douglas was still proposing a similar concept in the DARPA studies of 1974 –1975. Perko, therefore, encouraged McDonnell Douglas to team with Teledyne Ryan, because of that company's demonstrated expertise in designing low RCS drones. This teaming arrangement was made during early 1975.[188]

Program Formulation

During the summer of 1975, it had become apparent to Perko that at least two of the contractors he was working with—Lockheed and Northrop—were

*"Yehudi Lights" evolved from theoretical studies and experiments conducted beginning in the 1930s and continuing with full-scale flight tests using Grumman TBF and Consolidated B-24 testbeds during World War II.

on the verge of a revolutionary breakthrough in RCS reduction. Perko, therefore, met with Bob Moore,* who had replaced Kent Kresa as head of DARPA's Tactical Technology Office, and George Heilmeier, the director of DARPA, to determine the future direction of the program.[188] They decided at this point to take the program through flight test. Because of the almost unbelievable magnitude of the predicted breakthrough, it was felt that a full-scale flight demonstration would be necessary to make the results convincing. The program would start with a relatively short phase I effort culminating in a ground RCS evaluation of large scale models. Based on this evaluation, together with preliminary aerodynamics, flight control, and performance estimates from the phase I participants, one contractor would be selected to proceed with phase II, the construction and flight testing of two demonstration vehicles.[188]

Heilmeier supported this plan, and the next step was to obtain funding. Most DARPA programs at that time were based on a "34 - 33 - 33" cost split between DARPA, one of the armed services, and the prime contractor. DARPA retained program management control by contributing slightly more than one-third of the funding, while the other two participants each contributed slightly less. Most DARPA programs were too high risk for either the contractor or the service branch to want to fund in full; however, providing roughly one-third of the funding for such projects was often considered an acceptable investment for both the contractor and the service branch that was involved.

For the XST program, the estimated cost was around $36 million. The participating contractors had already expressed their willingness to share costs

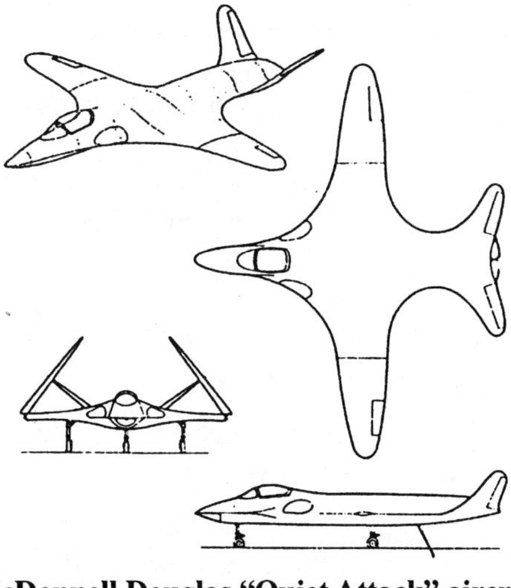

Fig. 10 McDonnell Douglas "Quiet Attack" aircraft for ONR.

*Moore had replaced Kent Kresa as head of DARPA's Tactical Technology Office. He later became director of air warfare, Office of the Director of Defense Research and Engineering (ODDR&E).

of approximately $10 million. However, Air Force support was less sure. The Air Force had just completed the lightweight fighter program, an "experimental prototyping" effort that was leading to a major production program for a fighter (the F-16) that certain elements in the Air Force did not want. As a result, there was some reluctance to undertake yet another advanced concept flight demonstration.[188, 234]

As a first step, Heilmeier approached Dr. Malcolm Currie, Director of Defense Research and Engineering (DDR&E) in the Office of the Secretary of Defense (OSD), to seek his assistance. Perko and Heilmeier presented a 15-slide briefing consisting of artists' drawings of all of the proposed aircraft concepts, a summary of the threat analysis, and a statement of the predicted breakthrough. Currie was supportive and suggested that Air Force funding be secured immediately. Currie arranged meetings that same day with Air Force Lt Gen Alton D. Slay, deputy chief of staff for research and development, and Gen David C. Jones, Air Force Chief of Staff. By the end of the day, Perko and Heilmeier had given their briefing three times and had been promised Air Force support for the program.[188]

Discussions were then held with Air Force Systems Command for approval to support the program from the RPV/Special Projects organization at ASD, where Perko had worked before his assignment to DARPA. Bill Elsner at that office was already providing technical support for Perko's studies. AFSC gave permission, with the proviso that the staff also would continue to perform their present duties.[188] Later, near the beginning of phase II, XST/Have Blue would have a dedicated project office. In addition to Elsner (who was effectively chief engineer), the Air Force team supporting XST grew to include Alan Atkins (observables), John Griffin (systems), Skip Hickey (flight controls/flying qualities), Ed Dow (aerodynamics), Jerry Forner (internal aerodynamics/propulsion), and Ralph Grimm (propulsion).[40, 234] Additional technical experts from the Massachusetts Institute of Technology and other research centers were brought in as consultants when required.[34, 197] Contracting was accomplished through the RPV office, as it had been during the earlier studies.

Phase I Source Selection

By August 1975, with funding and other necessary arrangements made, proposals were informally requested from the three participants. Proposals were to cover phases I and II. Phase II entailed developing and flight testing two demonstrator aircraft. The proposals stated signature goals as derived from the earlier study phase of the program and were to be based on roughly a one-third cost sharing by the contractor.

In their proposals, Lockheed and Northrop both submitted concepts that they predicted would meet or exceed the stated XST signature goals. The

McDonnell Douglas/Teledyne Ryan concept did not come close to achieving the goals, and proposed to use ECM to make up the difference. That was not the intent of the XST competition. After determining to the best extent possible that the Lockheed and Northrop proposals were credible, DARPA, therefore, selected these two contractors to proceed with phase I of the XST program.[188]

As the phase I proposals were received and reviewed, DARPA and the Office of the Director of Defense Research and Engineering discussed the implications and potential importance of the predicted breakthrough. These discussions resulted in a decision to protect the program at the Top Secret level. DARPA was not very well equipped to manage a program at this level, although Ken Perko had experience with such programs from his earlier assignment at the ASD Special Projects Office. Because that office was already providing technical and contracting support for Perko's study effort, DARPA managers thought it logical that ASD special projects office would assume an increasing managerial role in the coming phase of the XST program.[188]

XST Phase I

Contract Award

On November 1, 1975, Lockheed and Northrop were awarded contracts of approximately $1.5 million each to conduct phase I of the Experimental Survivable Testbed program.[126] Phase I would involve a four-month design and development effort, culminating in a competitive evaluation of full-scale RCS models at the Air Force Radar Target Scatter (RATSCAT) facility in New Mexico. Based on the results of phase I, the plan was to select one contractor for phase II, the development, construction, and flight testing of two demonstrator aircraft. According to Perko, two full contracts through flight testing were actually written at this time, with phase II as an option following phase I. It was understood that the option would only be exercised for one of the two contractors.[126, 188]

In addition to RCS testing, an increasing effort during phase I was devoted to ensuring that the XST was in fact a workable aircraft design. This effort included continuing refinement and analysis of the air vehicle design; low speed, high speed, and propulsion (inlet) wind tunnel testing; and initial flight control simulation.[126]

Lockheed Phase I Design Effort

At Lockheed, Dick Scherrer was chief preliminary designer, assisted by Ken Watson and others who were added to the team as the workload increased. Principal engineers were Alan Brown (low observables design), Denys

Overholser (low observables RCS theory and measurement), Ed Baldwin and Henry Combs (structures), Dick Cantrell (aerodynamics), Dave Campbell (propulsion), and Bob Loschke (flight controls).[175, 186, 204] Most were Skunk Works veterans with experience on the U-2 and SR-71. Scherrer, Watson, Brown, and Loschke came from the Lockheed California main plant and had been involved in the S-3 Viking program.[181, 204] Ed Martin also left his position as Lockheed's "white-world" science and engineering director, to rejoin the Skunk Works and work on XST.[151, 197] Because of the very tight four-month schedule to perform all of the engineering and model test activities, the design effort at the Skunk Works proceeded at an intense pace, and not without some shouting matches over the unusual design features necessary for signature control.

Early in phase I, the ECHO computer program was relied on heavily to perform rapid-turnaround assessment of design changes. According to Overholser, Scherrer typically submitted a new design to the computer group at the end of a workday. Overholser's team would punch the deck of input data cards for the new geometry in the early evening and perform a quick run (perhaps only one or two aspects) to check for any obvious mistakes in the geometry input. Errors, if found, were corrected, and then the team would submit the complete analysis job—a full set of elevations and aspects—to the computer and go home. They would pick up the results at 6 a.m. the next morning, check them over, and provide them to Dick Scherrer. Based on the results, Scherrer and his team would make changes and submit a new design iteration before they went home that same evening. Working in this way, 20 design iterations were accomplished during phase I of the XST program in not much more than 20 days.[186]

ECHO RCS predictions were supplemented and validated with test data. Small scale tests were conducted at Lockheed's indoor anechoic chamber.

One-third scale models were tested at McDonnell Douglas' Grey Butte Microwave Measurement Facility in December 1975, and again in early 1976, prior to fabricating the full-scale model for the "pole-off" against Northrop.[126, 175] As noted earlier, the XST's signature levels were too low for accurate measurement in most existing test facilities, so the team devoted considerable effort to developing special mounting devices and improved test techniques.*

"Pole-Off"

It was determined very early in phase I that the XST models would have a lower RCS than the styrofoam mounting poles that were in use at RATSCAT, the site for the competitive full-scale RCS tests. Government representatives, Lockheed, and Northrop convened at a special session to determine a solution. Lockheed provided primary design input for a low-signature pylon and Northrop for a mounting system that could present the model at all required aspects while maintaining the low signature of the support hardware. Lockheed was assigned to build the new 40-ft-high support system, primarily because the company seemed able to accomplish the task more rapidly.[188] An additional $187,000 was therefore authorized to Lockheed for this purpose.[88, 126, 128] The support system was completed and delivered to RATSCAT by January 1976 and was used to test both contractors' models.[188, 235]

According to both Ken Perko of DARPA and Denys Overholser of Lockheed, the RATSCAT staff did not at first take the program very seriously,

*Overholser, as Lockheed's primary radar engineer on the project, became heavily involved in this effort, and remains so through the present time. He now operates Denmar, Inc., a low observables consulting firm in Carson City, Nevada.

particularly when the new model support system was delivered.[186, 188] The XST team gained credibility, however, as tests began. From that point onward, the XST program received excellent support at RATSCAT. The competitive RCS tests were conducted through most of March 1976. Both the Lockheed and Northrop XST models demonstrated signatures that were orders of magnitude lower than any air vehicle previously designed.[235] Throughout most of the testing, the competing contractor teams and their models were kept in isolation from each other. Temporary quarters were set up so that each group would have access to the range but could be kept apart from the other team. However, after most of the testing was completed, each team was allowed to drive out on the range to view the other's model mounted on top of the 40-foot pylon.[186, 204]

Technical performance, risk, cost, and schedule of the two competitors were very close.[188, 235, 204] Consequently, choosing the contractor for phase II was somewhat subjective. Technically, the Lockheed entry had a slight edge.

Although both companies had developed special materials and edge construction techniques, the perception was that Lockheed Skunk Works had more experience with the use of these techniques on actual aircraft. The Skunk Works also had a proven record of accomplishing advanced, high-risk projects rapidly and under tight security. These factors provided confidence that the company could execute the XST program successfully.[35, 188, 235] Accordingly, in April 1976 Lockheed was announced the winner of the phase I competition.

At the same time, DARPA recognized that the Northrop XST team was an important center of expertise and urged the team to stay together. Shortly afterward, DARPA initiated design studies for a Battlefield Surveillance Aircraft, Experimental (BSAX).[204, 228] The BSAX studies led to the recently acknowledged Tacit Blue program for a very low observable battlefield surveillance aircraft.[73, 188] Tacit Blue was a very successful flight demonstration program that provided valuable data for the subsequent B-2 bomber program. Keeping two contractors active in the field of low observables was an important action that provided continued technological advances and maintained the pressure of competition as an incentive for maximum performance and innovation by both contractors.

XST PHASE II/HAVE BLUE

What was conceived as phase II of the XST program, and soon became the Have Blue program, was initiated on April 26, 1976, when Lockheed Advanced Development Projects was authorized to proceed with the design, construction, and flight testing of two demonstrator aircraft. Contract negotiations were completed in June 1976.[128] First flight was scheduled for December 1977.[126] The XST aircraft were not to be prototypes but rather technology demonstrators with three specific objectives[88]:

1) Validate the achievement of low observability, including very low RCS and infrared, visual, and acoustic signatures in flight with a manned aircraft.

2) Demonstrate that such an aircraft could have acceptable flying qualities and performance.

3) Demonstrate modeling capabilities and test techniques to accurately predict low observable characteristics of actual aircraft in flight.

XST was not intended to prove or test a specific weapon system design; the requirements for such a system had not yet been formulated. The intent was that if the program achieved the stated objectives, developing a very low observable military aircraft could then proceed with confidence.

By the initiation of phase II, several important things were beginning to happen. When the original phase I XST proposals were received in early fall of 1975, there had been some debate over whether the breakthroughs predicted by Lockheed and Northrop were real. As previously noted, it was determined that the claims appeared to be valid, and a decision was made that such a breakthrough should be immediately protected at a higher level than it had been up to that point.[188] The XST program classification had, therefore, changed to Top Secret early in phase I. Neither DARPA nor the contractor were well prepared to run a program at that level, however.[186, 188, 197] During the course of phase I, the security concept evolved, and by the initiation of phase II the Air Force Systems Command assumed management of the effort as a Covert Special Access program (although some of the original DARPA leadership remained involved).[188, 234] General program information—including even the fact that the program existed—was classified Secret/Special Access Required (Secret/SAR), and only the most critical technical data were Top Secret/SAR.*

At this time, the program and aircraft were named Have Blue and were thenceforth officially referred to only by that name. The XST designation was discontinued, and the participants from the earlier study phase of the program were led to assume the initiative had ceased.[188, 234, †] "Carve-out" actions eliminated Navy Plant Representative/Defense Contract Audit Service (NAVPRO/DCAS) cognizance of the program at Lockheed, and a "black" contracting and procurement system was established.[88, 128]

The ASD staff already supporting XST became the core of a dedicated project office in a secure vault at ASD Headquarters, Wright–Patterson Air Force Base, under the leadership of Maj John K. (Jack) Twigg.[234] Also at this time, with the increased confidence provided by the results of the phase I "pole-off," Currie (DDR&E), together with Gen Jones (Air Force chief of staff) and other senior leaders who were aware of the program, began to formulate a strategy for developing and fielding the technology in operational weapons systems. The group secured the support of several key government and military officials and obtained funding to start the U.S. Air Force's formal low observables effort. Starting in the summer of 1976, Air Force Headquarters began to conduct preliminary studies to determine how the XST breakthrough could best apply to operational systems.[115]

---

*Confidential, Secret, and Top Secret are levels of classification that do not in themselves restrict access to a particular set of individuals. The Special Access classification further restricts information at each level to those individuals who have been "briefed in" to a particular program.

†Perko stated that he used the level of other companies' progress in low observables to gauge how well the Have Blue program security was working. As an example, General Dynamics continued to emphasize countermeasures as a necessary complement to RCS reduction, noting in a 1977 study that, "At some value of airframe RCS the ECM antennas will become the dominant RCS source of the airframe-with-antenna combination. Further reduction of airframe RCS through shaping or treatment will yield very little change in total RCS.... It appears unlikely that an operationally useful super stealthy airframe able to avoid detection by the SA-X threat passively (without ECM and associated antennas) will be physically possible."[6] As further evidence of the effectiveness of the security effort, the Have Blue program remained unknown to the public until more than 10 years after its completion. Some contemporary press reports related to Have Blue and the early development of low observables technology are provided in Appendix D for historical interest.

In June 1977 a small project office was established for this purpose at Air Force Headquarters in the Pentagon, with an initial staff of five officers.[25] The magnitude and importance of the technological breakthrough that was involved "resulted in joint decisions by DARPA, DDR&E, and the Air Force to protect the low observables program as Special Access Required. The requirement was to protect the existence and nature of the program, the level of commitment, and the technology details."[234] Minimum staffs and short, direct lines of communication were used both in the interest of security and also to facilitate rapid, informed decision making. The ultimate objective of the effort was to develop a strong and uncompromised U.S. lead in the field of low observables to help offset the superior numbers and formidable radar-based air defense systems possessed by the Soviet Union and its allies.[115, 188, 234]

Although for security reasons Have Blue procurement and contracting was done outside of normal channels, the program still operated with far less latitude than the subsequent Senior Trend (F-117) program. Have Blue also did not at first possess the clout or the money that Senior Trend would later.[25] According to Twigg,

> It is relevant to note that in the early stages of Have Blue the Air Force senior level interest was at best minimal. Other challenging priorities at the time on the F-15 and B-1 programs, among others, were the clear focus. Inputs and coordination through DARPA and DDR&E to [Air Force] Chief of Staff Gen David C. Jones resulted in his direction to allocate AF funds in support of the Have Blue project [as described earlier]. Clearly Air Force interest increased as the program achieved major milestones in pole RCS tests, ground tests, and later, flight tests.... A Major from Tactical Air Command named Joe Ralston* was a key regular interface [between the Pentagon low observables project office and the Have Blue program] tracking progress and providing advice and support.[234]

Rapid, direct communication was a key feature characterizing both the Have Blue and Senior Trend programs. Senior officials, including (from 1977 onward) Director of Defense Research and Engineering William J. Perry and Secretary of Defense Harold Brown, were kept apprised of the progress of the Have Blue program. This helped to establish the stable, top-level support that later played a key role in the success of Senior Trend.

The Have Blue program management also valued close cooperation between the government and contractor teams, real-time decision making, an absence of adversarial relationships, and a "try it–fix it–try it again" approach

---

* Gen Ralston later became commander of the U.S. Air Force's Air Combat Command (ACC) and at the time of publication is vice chairman of the Joint Chiefs of Staff. As a major in 1977, he was one of the original five members of the Air Force low observables office in the Pentagon.

with emphasis on solving problems rather than assigning blame. The ASD team met at contractor facilities where the action was. For this reason, the team members were selected not only for their technical abilities but for their ability to think on their feet and endure very demanding travel schedules. The average age of the program office personnel was approximately 35.[188, 234] The government team was small, but each member was highly qualified. Most were multiskilled and all were highly committed to the principles already mentioned. There were "no prima donnas or glory hounds."[233] The same criteria generally applied to the contractor's selection of people for the project. Both at the ASD office and at the Skunk Works, physical working conditions were far from comfortable, but the people involved developed a strong sense of teamwork and a high degree of motivation.[40, 186, 234]

The participants enforced a multidisciplinary, systems engineering approach. Unlike traditional aircraft developments, "the signature engineers had equal status with the aerodynamics/stability and control specialists in driving the aircraft design solution."[234] In fact, it has been said that the aerodynamicists and other engineers in "traditional" disciplines were actually subordinate to the signature engineers. On the contractor side, this was a cultural change that was only accomplished with a great deal of persistence from those who understood how a rigorous and comprehensive effort was necessary to achieve any benefit from signature reduction. Without this disciplined approach, a very low observable aircraft could not have been built. The basic physics of the problem, chiefly that detection range varies only as the fourth root of RCS, simply did not leave much room for compromise.

Aircraft Development and Fabrication

Lockheed's team grew in preparation for the flight demonstration phase of the program. Norm Nelson became Lockheed's Program Manager for Have Blue and reported directly to Skunk Works General Manager Ben Rich. Nelson had previously been on a special assignment aboard the research ship *Glomar Explorer.*[151] Bob Murphy joined the team as director of manufacturing.[175] Alan Brown, the principal low observables engineer for XST, became chief engineer for Have Blue. During the summer of 1976 Dick Scherrer, chief preliminary designer, suffered a stroke that ended his involvement with the Have Blue program.[204, *]

Before and during construction of the actual Have Blue aircraft, tests conducted included roughly 1000 h in low- and high-speed wind tunnels, 500 h

---

*Scherrer returned to the Skunk Works approximately one year later and worked on a preliminary design for a low observable reconnaisance aircraft with efficient, conventional airfoils and a four-spike configuration, i.e., one in which all planform edges were oriented in only two directions, parallel with either the port or starboard wing leading edge. The Skunk Works rejected this concept, and in 1979 Scherrer left Lockheed to join the design team at Northrop for the Advanced Technology Bomber (ATB), which became the B-2.[204]

in a propulsion wind tunnel, 88 test days at McDonnell Douglas' Gray Butte RCS measurement range with a one-third scale model, and 40 test days at RATSCAT with a full scale RCS model, beyond what had already been performed during XST phase I. Other tasks included structural testing, mockup, continued flight control system development and simulation, and "iron bird" functional mockup testing.[126]

Two Have Blue aircraft were built. Fabrication of Have Blue aircraft No. 1 began in July 1976.[126, 197] As construction got underway, Alan Brown spent a lot of time on the factory floor answering questions and ensuring that all aspects of aircraft design and fabrication were conducted in a manner consistent with signature control.[34, 197] The general arrangement and principal characteristics of the Have Blue aircraft are illustrated in Fig. 11 and Table 1.

As a cost- and time-saving measure, only the second Have Blue aircraft had full low observables treatments. The first aircraft was for handling qualities and flight performance evaluation only, while the second aircraft was for low observables testing. In keeping with its particular purpose, Have Blue No. 1 was equipped with a large test instrumentation noseboom and a drag chute mounted in an external box on top of the aft fuselage. Neither of these distinctively non-low observable features was on the second aircraft. Another difference between the two aircraft was that Have Blue No. 1 did not include the nosewheel steering which was incorporated on Have Blue No. 2 to improve its ground handling qualities.

SUBSYSTEMS AND EQUIPMENT

The Have Blue demonstrators were as small and simple as possible to minimize cost. No mission equipment was included, the cockpit was unpressurized,

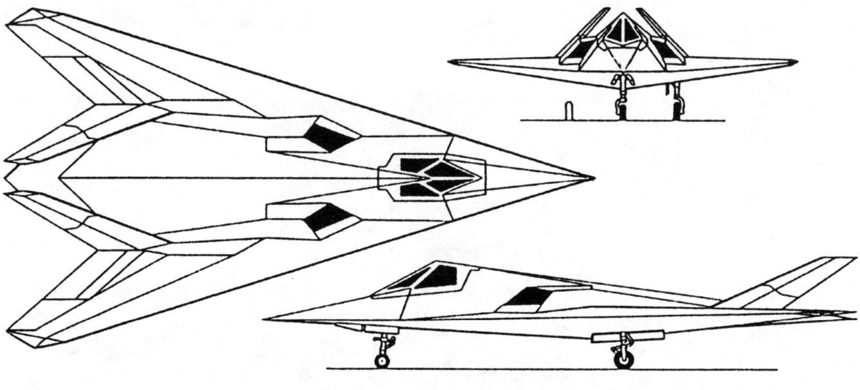

Fig. 11 Have Blue three-view.[90] (Courtesy of Jane's Information Group.)

Table 1 Have Blue aircraft characteristics[104]

Length	38' (w/o noseboom)
Wingspan	22' 6"
Height	7' 6"
Wing area	386 sq ft
Wing sweep	72.5 deg
Empty weight	8950 lb
Maximum fuel weight	3500 lb
Payload	none
Maximum takeoff weight	12,500 lb
Range/endurance	1 h endurance
Maximum speed	0.8 Mach
Propulsion	2 X General Electric J85-GE-4A 2850 lb thrust each

and the aircraft had no flaps or speedbrakes and no air-to-air refueling capability. Conventional materials were used in the construction of both aircraft with RAM applied over the aluminum structural skin, and only to Have Blue No. 2. Maximum use was made of off-the-shelf components. Maj Twigg proved to be extremely skillful at obtaining such items covertly.[197] According to Twigg,

> One of my key jobs...was to secure components from many sources without compromising security. This proved to be extremely challenging and for Air Force elements I had to occasionally invoke "need is classified. Please call Lt Gen Sylvester (ASD Commander) if you wish." I do not think anyone did! Some key components were F-16 side stick

> controller, flight control computer, and seat. General Abrahamson [as head of the F-16 SPO at the time] was extremely cooperative.[234]

Other off-the-shelf items included F-5 landing gear and proven General Electric J85-GE-4A engines similar to those used on the Rockwell T-2C, Cessna A-37, and Northrop T-38 and F-5 aircraft. The six engines acquired for the Have Blue program came directly from the Navy's stock for the T-2C. "For [the engines] I arranged some horse trading of Air Force components through AFLC [the Air Force Logistics Command] for the Navy J85-GE-4A engines at no cost to the program," recalls Twigg.[234]

Flight Control System

The Have Blue demonstrators were designed to be statically unstable about the pitch axis.[101, 175] This was not an inevitable consequence of their faceted shape. Almost any shape will be stable if the center of gravity is far enough forward. A very low observable aircraft could have been designed with positive stability about all axes. However, the designers chose to make use of emerging automatic control technology to relax some traditional stability and control constraints. Most notably, an aft center of gravity position was used to minimize performance penalties (low maximum trimmed lift, high drag-due-to-lift, and high trim drag) associated with high wing sweep, faceted airfoils, and lack of a horizontal tail surface. The aft center of gravity made the aircraft unstable in pitch. "Artificial stability" was then provided by a quadruplex "fly-by-wire" control system modified from the F-16. Fly-by-wire means the *commands* to the flight control actuators were electrical rather than mechanical. However, the *power* for the actuators on Have Blue—and nearly all other fly-by-wire aircraft flown to date—was hydraulic. At the time of the Have Blue program, the F-16 was not yet operational but the control system had flown on the YF-16 Lightweight Fighter demonstrators during 1974. Have Blue, like the F-16, was only designed to be unstable in the pitch axis. However, the flight control system used closed-loop active control in all axes to provide stability and command augmentation.

Because the YF-16 was a demonstrator for what was hoped and in fact turned out to be the world's first statically unstable fly-by-wire operational fighter, its control system had been designed with as much flexibility as possible to allow rapid development and fine tuning of its control laws. This valuable feature allowed many flying qualities problems to be solved quickly and easily during the YF-16's flight tests, and also allowed the same basic flight control system to be adapted readily to the production of F-16 and other aircraft.[13] Have Blue used the flight control computers and side stick controller of the YF-16 but required new control laws, new air data interfaces, and a new failure detection/redundancy management scheme.[175] New control laws were needed because of the different stability characteristics of Have Blue relative to the YF-16. Have Blue also had a completely different air data system and arrangement of control surfaces than the YF-16; thus the interfaces and scheme for dealing with sensor or actuator failures had to be revised. The successful accomplishment of these modifications is a testimony to the flexibility designed into the YF-16's flight control system.

In any statically unstable aircraft, the flight control computer depends on input from the air data system to make continuous attitude corrections. A conventional aircraft would simply be equipped with a sufficient number of probes and vanes to provide unambiguous information on airspeed, altitude, pitch, and yaw. Have Blue's need for RCS reduction prohibited the usual array of probes, so a unique system had to be devised. This consisted of three total pressure ports, one at the point of the nose and two on the canopy centerline post, plus three static pressure ports installed flush with the aircraft's surface at various locations on the forward body. Initially, wind tunnel tests determined the air data correction factors and final adjustments were made based on comparison of the noseboom data from Have Blue No. 1 with data from the pressure sensors.[175] Static and total pressures aided altitude and airspeed calculations in the usual manner. However, the number and location of the air data sensors were not adequate for measuring pitch α and yaw β angles reliably and accurately. Instead, α and β were derived from inertial data using gyros and accelerometers.[34, 175] Because of Have Blue's mission as a test aircraft, it contained no deicing provisions for the air data system. The aircraft would only be flown under safe, nonicing conditions.

Flight Test

Have Blue No. 1 was completed in November 1977 and transported by C-5 to the test site.[88, 197] Final assembly was completed by Lockheed managers due to an aerospace machinists' strike, which had started a few weeks earlier.[34, 197] In spite of this setback, the first flight took off on schedule on December 1, 1977, with Lockheed Pilot Bill Park at the controls.* In spite of a strong headwind, the takeoff run was long and the climb was sluggish.[197] Nevertheless, the flight was accomplished successfully. In the following months, Have Blue No. 1 conducted functional checkout, performance, and flying qualities tests.

Very tight security accompanied all flight testing. Flights were only made during the first hour or so after dawn each day to minimize the chances of an unauthorized sighting. Have Blue personnel flew to the test site from Las

*Park was a veteran Skunk Works test pilot who had previously survived two A-12 crashes. One of them involved an aileron servo malfunction, which caused an uncommanded large amplitude roll at low altitude; the other resulted from a collision with a D-21 reconnaissance drone during an attempted launch of the drone from the back of the A-12.[197]

Vegas every week, arriving Monday morning and leaving Friday evening. This gave them four mornings per week to fly. Most other site personnel commuted daily. Have Blue would be finished with its flight and packed away in the hangar before most of the other site personnel arrived each morning.[34] Those who were at the site on the mornings of testing and not cleared into Have Blue had to be indoors with window shades closed whenever Have Blue left the hangar. Furthermore, the aircraft was kept in the hangar during overflights of any known or suspected Soviet reconnaissance satellites.[34, 229]

PERFORMANCE

Limited performance, evident from the first takeoff run as noted in the previous section, resulted from a combination of low thrust-to-weight ratio and low lift-to-drag ratio. The thrust-to-weight ratio would have been between 0.46 and 0.62 depending on fuel load and based on the rated thrust of the engines. However, the aircraft lost additional thrust due to the inlet grids and the slotlike exhaust nozzle. The maximum lift-to-drag ratio of Have Blue was approximately 7.5. For comparison, typical fighter-type aircraft have thrust-to-weight ratios between 0.75 and 1.0, and lift-to-drag ratios between 9 and 12. A comparison of Have Blue's lift-to-drag ratio with that of other military aircraft is illustrated in Fig. 12.[88] [(Span)^2/(wetted area) is used for the horizontal axis because this is a measure of what lift-to-drag should be achievable. All other things being equal, lift-to-drag should track as the square root of this quantity.]

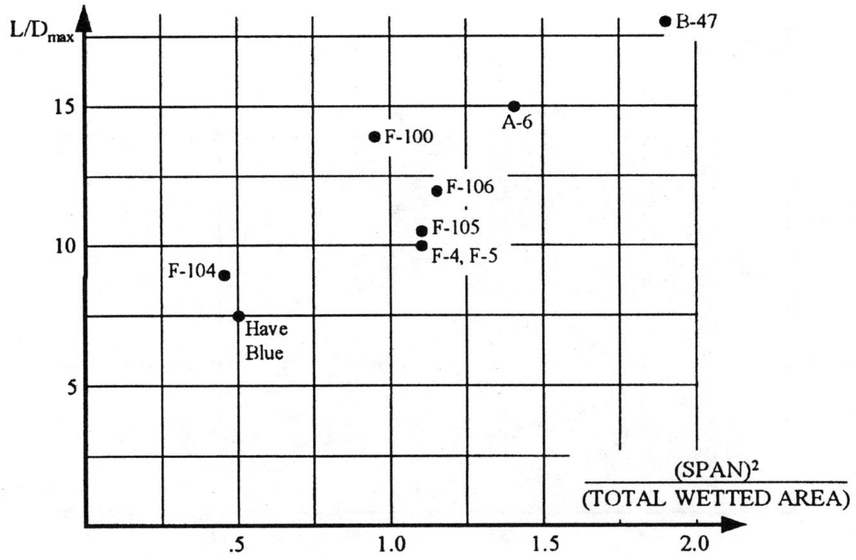

Fig. 12 Lift-to-drag ratio comparison of Have Blue and other aircraft.[88]

The faceted shape contributed only minimally to the low lift-to-drag ratio. Much more important were the high leading edge sweep and small span (low aspect ratio) of the wing. Neither of those features is conducive to efficient lift production, particularly at low airspeeds. The limitations of the basic wing design coupled with the lack of high lift devices to generate added lift or of a separate horizontal tail to help trim the airplane also resulted in very high takeoff and landing speeds. Because of these and other design compromises, Have Blue operated within a relatively limited flight envelope as show in Figs. 13 and 14. Maximum altitude was 25,000 ft and the limit load factors were +3 and $-1g$.[34, 88]

Flying Qualities

The peculiar aerodynamic shape and unique propulsion installation of Have Blue produced unusual stability characteristics. In the pitch axis, the high wing leading edge sweep (72.5 deg) was known to cause large nose-up pitching moments at moderate to high angles of attack. Tests, including indoor free-flight model tests conducted at the Skunk Works' assembly building, determined that pitch departure would occur if the aircraft exceeded 17 deg angle of attack.[34] To prevent this from happening to the full-sized aircraft the team devised a crude angle-of-attack limiter. If the aircraft exceeded approximately 13 deg (selected to provide a 4 deg margin below the pitch departure condition) a section of the exhaust deck trailing edge (the "platypus") would deflect downward automatically to provide additional nose-down control. After

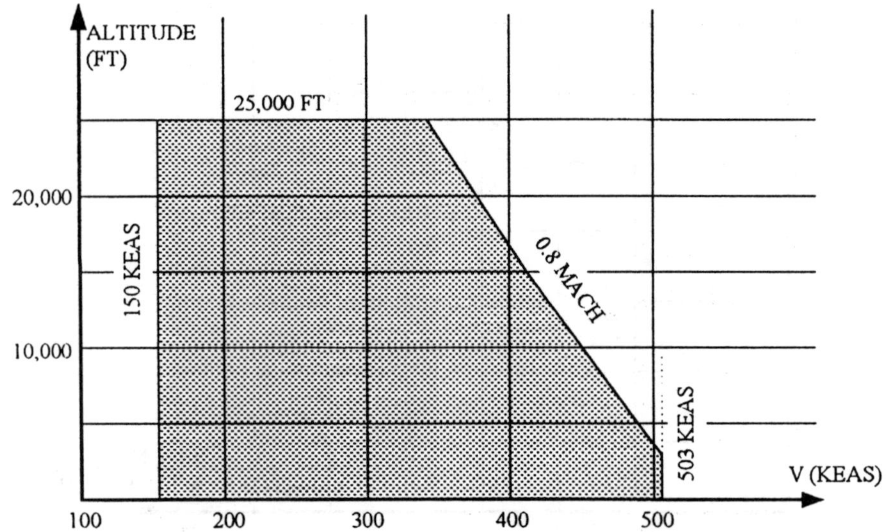

Fig. 13 Approximate Have Blue flight envelope.[88]

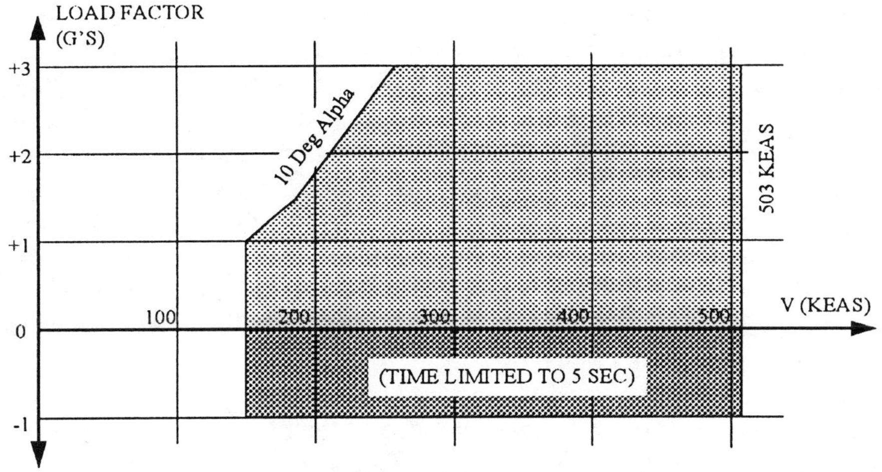

Fig. 14 Have Blue V-N diagram.[88]

recovery to another lower threshold, the platypus flap would return automatically to its normal position.[34, 101] According to Skip Hickey, Air Force Flying Qualities Engineer for Have Blue, "This scheme was not optimum and led to control difficulties during slow speed landings. It was not used on the F-117A."[101]

In the lateral/directional axes (yaw and roll), the highly swept wing produced a large amount of dihedral effect (lateral stability), which increased with increasing angle of attack. Too much dihedral effect can lead to wing rocking, or "dutch roll." The strong inward cant of the vertical tails caused the rudders to produce pitching and rolling moments as well as yawing moments when deflected. These effects, while challenging to overcome, had been predicted based on wind tunnel tests. Through analysis and simulation, control laws had been developed to provide "level 1" flying qualities* within the flight envelope defined for Have Blue. However, according to Hickey,

> The aerodynamic characteristics...resulted in the proverbial flying qualities "cliff." When flying a relaxed stability airplane, everything feels good until a control surface deflection limit, and/or deflection rate limit, is reached.... Flight outside the flight envelope resulted in the same problems that would be encountered with any relaxed static stability aircraft.[101]

Before starting the actual flight test program, extensive piloted simulations were performed to assess the vehicle's behavior and provide experience for the pilots in dealing with a variety of foreseeable failure states.

---

*A standard definition of level 1 flying qualities is "desired performance attainable with a tolerable workload"; level 2 corresponds to "adequate performance attainable with a tolerable workload"; level 3 to "adequate performance not attainable with a tolerable workload"; and the next level, not assigned a number, is loss of control.

The full effects of the wide exhaust nozzles on the aircraft's lateral/directional stability had not been completely determined prior to flight testing. This would have required wind tunnel testing of the complete configuration with properly scaled flow through the inlets and exhaust nozzles, which was not done in the rapid, low-cost Have Blue program. Possibly as a result of these thrust effects, there were some discrepancies relative to the predicted lateral/directional characteristics of the aircraft. The directional stability was less than predicted, and the aircraft was directionally unstable above 0.65 Mach. This was successfully compensated for with the yaw gain in the control system. Side force due to side slip was also less than predicted, as was roll damping below 0.3 Mach.[175]

A very unusual characteristic occurred because the net thrust vector from each engine was "toed-in" somewhat, due to the highly swept plane of the exhaust nozzles. As long as both engines were operating, their respective side force components from the nozzles would cancel each other; but if one engine failed the net effect was roll and yaw *toward* the good engine, rather than away from it, which would normally be the case. This, combined with the roll/yaw coupling generated by the rudders, "resulted in an iterative process to trim roll and yaw. Though highly entertaining, the pilots found it too time consuming."[101]

Additionally, exhaust heat caused warpage of the platypus nozzle surface. According to Hickey, "a special team headed by [Henry] Combs of ADP rebuilt the aft fuselage of Have Blue No. 2 to address this problem."[101] The subsequent F-117A nozzle used a different design incorporating a shingled platypus upper deck in which panels could slide over one another to accommodate nonuniform thermal expansion.[101, 175, 229]

Flight Test Summary

From December 1977 through early 1978, 36 test sorties were accomplished with Have Blue No. 1, all of them for functional checkout, flying qualities, and performance evaluation.[88] The first few flights were piloted by Lockheed's Bill Park after which an Air Force pilot, Maj Ken Dyson, began to fly the aircraft as well.[50]

Have Blue No. 1 was lost on May 4, 1978. Over the previous several flights leading up to the accident, Park had been progressively lowering his landing speed by small increments. The platypus was not intended to be used for landing approach; landings were to be made at a lower angle of attack than the angle at which the platypus was designed to operate. For this reason, nobody gave much thought to the possible consequences of platypus operation during a landing approach. However, as Park attempted to land at a lower speed, and correspondingly a higher angle of attack, he reached the threshold angle of attack for platypus operation just before touchdown on that particular morning. The sudden nose-down pitching moment caused by the automatic platypus deflection resulted in a very hard ground contact (this sequence of events was referred to in the official Air Force briefing as "pilot–platypus interaction").[88] Park lifted off again to regain control and retracted the landing gear for a go-around. However, the gear had been damaged and would not extend for a subsequent landing attempt. A decision was made by the pilot and the test personnel on the ground to eject rather than attempt a wheels-up landing. Park sustained injuries resulting from the ejection that ended his flying career. He stayed at Lockheed ADP as director of flight operations until he retired in 1989.[88, 101, 229]

Have Blue No. 2 made its first flight on July 20, 1978, and completed 52 sorties.[234] Ten of these were for functional checkout and performance evaluation and the remaining 42 were for low observables testing.[88] A large number of signature tests were conducted, and Have Blue successfully demonstrated its design goals and convincingly showed that extremely low signatures could be achieved in flight by a manned aircraft. Agreement with predictions was good, providing confidence in predictive methods (both analysis and ground test) in the design of future systems. All flights with this aircraft were piloted by Maj Dyson, who later became Rockwell's chief test pilot for the X-31 enhanced maneuverability flight demonstrator aircraft.

Have Blue No. 2 was lost on July 11, 1979. An engine exhaust duct weld failed, allowing hot exhaust gases to leak into the engine compartment. This caused a failure of the first (of two) hydraulic systems providing power to various subsystems on the aircraft, including the flight control actuators. The pilot attempted to return to the base, but the second hydraulic system soon failed in a similar manner to the first one. This resulted in complete loss of aircraft control. The pilot ejected safely.[88] When Have Blue No. 2 was lost, the program was within two or three sorties of planned completion.[88, 115]

XST/Have Blue Program Summary

The Have Blue program officially ended in December 1979. Have Blue program duration was 44 months from phase II go-ahead on April 26, 1976, to completion in December 1979. The average number of Lockheed engineers was 66, including 42 designers and 24 supporting engineers in various disciplines. The peak number of engineers on the project was 200 from the fourth through the ninth month. The peak total of contractor personnel was 400, which was reached 13 months after program go-ahead. The first flight (December 1, 1977) was 20 months from contract go-ahead.[88, 126, 128]

The Have Blue technology demonstration program was extremely successful in that it accomplished all of its technical objectives. Signature goals were essentially met. Comparison of predicted signatures, with those actually achieved in flight, gave the designers of future low observable aircraft confidence in analytical prediction methods and ground test techniques. Flying qualities and performance testing gave insight into what the performance limitations were and what kind of military missions a very low observable aircraft might be capable of performing. According to former Air Force Have Blue Program Manager John K. Twigg, "A top level summary could express the outcome as a 20–30% reduction in aerodynamic efficiency compared to standard aircraft," in exchange for an orders-of-magnitude reduction in RCS, as well as significant reductions in infrared signature. This offered the potential for a vast improvement in survivability and overall mission effectiveness.[234]

In context of Ken Perko's original questions posed in 1974, the Have Blue program demonstrated that a manned aircraft could achieve signatures low enough to conceivably perform a tactical mission without being detected. The next step would be to determine how best to apply this unique capability, and to integrate the additional features to enable the aircraft to perform a useful mission.

Cost

The total program, including the preliminary studies, Lockheed cost sharing, government test support and other "miscellaneous" costs was $54.6 million in then-year dollars. (A detailed breakdown is provided in Appendix G.) Although the Have Blue program was initially underfunded, a cooperative approach involving a combination of contractor and government funding was successfully applied to achieve all original objectives.

Have Blue program cost was very reasonable compared to other demonstrator programs involving all-new airframes or substantially new airframe components. Some demonstration programs of comparable scope have achieved similar costs; few if any have done better, and many have done a lot worse.

OBSERVATIONS AND LESSONS LEARNED

Needs Identified for Follow–On Development

Have Blue identified several specific design issues that would have to be addressed in the follow-on development of any very low observable weapon system.

First, fabricating the aircraft and preparing for flight in a low-RCS condition required tremendous attention to detail. Doors and access panels had to be sealed with metallic tape and then covered over with RAM "butter," every time they were opened. The final step before a flight was usually waiting for the butter to dry.[175] Obviously, taping and buttering could not be done with the landing gear doors, so it was particularly critical that these fit perfectly with the surrounding surface. On Have Blue, this was not provided for adequately. The gear doors did not have an independent closing mechanism but instead were pulled up mechanically by the action of the retracting landing gear. This did not suffice to keep the doors flush at all times. Continual, positive closing forces would be needed for all doors. Furthermore, all door edges would have to be swept at the same angle as the other planform edges of the aircraft rather than aligning straight across the aircraft, as was the case with several of the doors and seams on Have Blue. Alignment with the aircraft's planform edges would be necessary to achieve the lowest possible RCS levels and to minimize chances of a serious degradation due to any door closing problems that might occur.[34]

Second, Have Blue had avoided the problem of developing low observable air data probes by relying on a small number of total pressure ports and static sensors mounted where they would not disrupt the basic contour of the aircraft, together with gyros and accelerometers that were mounted internally to the aircraft. The data provided to the flight control computers were adequate for basic flight control but probably not for tasks such as precision weapon delivery. Furthermore, scaling considerations indicated that the inertial scheme would not be adequate for a larger, mission-sized aircraft.[34] The problem of developing a low observable, fully functional pilot/static/α/ß probe would have to be solved. For accuracy, the probes would need to be in a conventional nose-mounted position, making signature control critical. To provide redundancy, multiple probes would be needed. Icing protection would also be required. As will be discussed later, the air data probes would prove to be one of the major challenges in the subsequent F-117A development.

Third, the automatic, two-position platypus flap provided less than optimum control characteristics. It automatically deflected at a prescribed angle of attack and retracted when the aircraft recovered to a certain point without the pilot commanding it or even necessarily knowing when it would operate. This could lead to very difficult situations for the pilot to control and was a contributing cause of the landing accident that resulted in the loss of Have Blue No. 1.

These lessons were learned by flying Have Blue. It was very worthwhile to learn them before undertaking full-scale development of an operational military aircraft. Other lessons would still have to be learned during the F-117A's full-scale development flight test phase. However, those would focus on integrating various features necessary to perform an operational mission. Have Blue provided the critical first experience building and operating a very low observable aircraft, allowing the development of the F-117A to proceed with a high degree of confidence in the basic vehicle design concept.

TECHNOLOGICAL BREAKTHROUGHS

Technology does not always proceed predictably through orderly stages from theory to phenomena observed in laboratories to applied ground demonstrations to flight demonstration and so forth. For instance, electromagnetic scattering theory evolved for more than a century, but when it reached the stage of usefulness in aircraft design and was actually understood by aircraft designers, it rapidly transformed the state of the art. Hit-and-miss, intuitive efforts at aircraft shaping progressed to a much more systematic approach that yielded orders-of-magnitude improvements in performance. Have Blue VLO design was completed in a very short development period by refining the aircraft shape to provide acceptable performance and handling qualities, by incorporating radar absorbing materials (which had been under development for decades), and by devoting special attention to details (such as engine inlets, the canopy, and the air data system).

However, it must be remembered that there was already a broad-based effort underway in the United States to develop RCS prediction and reduction techniques. A considerable effort in the field of RCS prediction was made during the latter half of the 1960s. Although Dr. P. I. Ufimtsev is widely credited with providing the key to Lockheed's breakthrough, the work that took place both before and after the "discovery" of Ufimtsev's theory was critically important to developing useful RCS prediction capabilities at both Northrop and Lockheed. This work was performed either on government contracts or independently to improve a company's competitive position for anticipated future programs. It is important for the government to invest in scientific research and advanced development programs to stimulate such progress.

The company that most successfully exploited the theoretical breakthroughs in RCS prediction had not even been involved in the original DARPA program because it was not in the fighter business at that time. As a result, Lockheed was almost left out of the DARPA/U.S. Air Force low observables effort altogether. The lesson is don't ignore centers of excellence. Lockheed Skunk Works was a center of excellence even if the company had not built any state-of-the-art fighters recently. According to Skip Hickey,

> The key engineering team leaders had worked on the F-104, U-2, A-12, SR-71, and who knows what else. Furthermore, Bob Loschke [the flight control engineer] had been brought on board from Lockheed Corporate [outside of the Skunk Works] and had been working on the S-3 Viking. The key people in management and engineering positions were *very experienced* with high-risk prototype programs in high-performance aircraft.[101]

To be convincing, technology demonstrations should be accomplished with clearly defined goals and/or competition between multiple contractors. From the initial studies through the final flight test phase, the XST/Have Blue program had very clear goals. Success or failure was not ambiguous. Competition was pursued through wind tunnel and RCS model testing. Both contractors demonstrated an outstanding capability during this ground test phase, and it was likely that either one could have successfully accomplished the flight demonstration.[188, 234] Because of funding constraints, only one contractor proceeded to the flight test phase of the XST/Have Blue program.

However, DARPA chose the second-place team (Northrop) to work on another program as soon as possible. This led to the Tacit Blue demonstration program, which worked other areas of low observables technology while Lockheed was occupied in developing the F-117A. Subsequent programs were able to benefit from competition between these two companies and eventually with other companies that developed low observables design capabilities. The ultimate result was a much broader experience and technology base than if a single company had become the U.S. Air Force's "sole source" for stealth aircraft.

Role of Technology Demonstrators

Appropriate demonstrations of new technology should in certain cases occur before proceeding with the design, development, and acquisition of an operational system that applies the new technology. Have Blue's initial subscale phase I ground testing indicated the potential for a significant breakthrough in RCS reduction. Flight testing of an inexpensive demonstrator validated that this breakthrough could be achieved with a manned aircraft, and only at that point was a weapon system acquisition program initiated.

Although the first phase of the XST program was limited to ground testing, the intent from the beginning was to demonstrate the predicted breakthrough in flight tests. Contracts awarded at the beginning of phase I actually included the flight test phase as an option, with the understanding that the option would be exercised with the winning contractor.[188] This helped to ensure that the demonstration would be sufficiently convincing to offer a real prospect of transitioning the technology into a practical weapon system with realistic operational utility.

Development and Testing

Small-scale, minimum-cost aircraft can be valuable in demonstrating specific technologies, but their limitations and scaling parameters must be understood. Have Blue demonstration aircraft predictably had marginal performance, a very limited flight envelope, and less than optimum flying qualities outside of that envelope. Risks were reduced as much as possible through extensive ground testing and simulation, including simulation of several foreseeable failure states.

The approach of tailoring each aircraft to a particular phase of the test program speeded up fabrication and reduced cost (because aircraft No. 1 did not have to be treated for low observables), but it also increased the risk that loss of an airplane would prevent accomplishment of all demonstration objectives. Fortunately, each Have Blue aircraft accomplished its demonstration goals before crashing. An alternative approach was taken by General Dynamics in the Lightweight Fighter program; the two YF-16s were nearly identical and so either one could be used for almost any test.[13] The best approach depends on program priorities and available budget. If the aim is to accomplish rapid coverage of a huge number of test points, then all demonstrators used should be as common as possible. On the other hand, if the aim is to demonstrate a unique, highly specialized new technology, as in the case of Have Blue, it may be more practical to tailor individual aircraft to specific purposes.

Another important tradeoff in a test aircraft is whether to incorporate capabilities such as air refueling. There should be no unnecessary complexity, but on the other hand air-to-air refueling capability can improve the productivity of a test program. Again, the correct choice depends on program priorities, and air refueling should be strongly considered if a fast-paced flight test program is desired. In the case of Have Blue, there were several good reasons not to include air refueling capability. A low observable air refueling capability would have had to be developed with additional cost and further security risk because air refueling crews would have to gain clearance for the program. Furthermore, the security procedures at the test site dictated that flying would only be performed for approximately an hour at a time, and the aircraft could fly that long without the need for air refueling.[34]

Six engines were secured for the Have Blue program, resulting in a ratio of three engines for every two-engine aircraft—a 50% margin of available engines relative to those needed at any given time. This ratio is typical and appears to have been maintained through the F-117A development and production programs as well, even though both Have Blue and F-117A engines were essentially off-the-shelf. If a program involves new engine developments, the ratio of engines available to engines required should be even more generous to allow for difficulties that almost always occur. Lack of engines is one of the most common causes of program delays or even outright failure,[106] and provisions should always be made to ensure that an adequate number of engines will be available. Even if the engines themselves are existing and in production, engine/airframe integration issues can be critical.

Have Blue No. 2 crashed because a single failure to an exhaust duct weld led to the loss of both hydraulic systems. "As is usual design practice, subsystems should be designed such that a single failure does not result in loss of control," notes Skip Hickey.[101] This means not only having redundant subsystems but locating the backups so that no single failure of any reasonable likelihood can disable both the primary and the backup of any flight-critical system. Exhaust gas leaks and other problems related to the integrity of the tailpipe would continue throughout the development of the F-117A. However, the hydraulic system layout of the F-117A was redesigned to ensure that a single exhaust leak could not cause loss of all hydraulic power to the flight controls. "We relearned this lesson!" [101]

PROGRAM MANAGEMENT

Both the government and the contractor teams operated with small staffs of highly qualified, highly motivated personnel. Excellent teamwork between government and contractor marked the Have Blue program. Jack Twigg recalled, "The reason we could work as fast as we did was trust on both sides, which was backed up with integrity." [235] An example of this trust is the way company and government funding was handled. Government fiscal years start on October 1, whereas the companies worked to the standard calendar year. During the Have Blue program, as well as the later Tacit Blue program with Northrop, both DARPA and the U.S. Air Force low observables effort as a whole were in a high-growth mode accompanied by confidence that next year's funding would be higher than the current year's. For this reason, the government (Tom Higgins at ASD was the contracting officer at this time) allowed and encouraged the contractor to spend the entire year's allocation of company funds in the first nine months of the calendar year, knowing that at that point the government funding for the following fiscal year would become available.[188]

This is not to say that disputes never occurred. They did, particularly with regard to who would have to pay for what.[128, 188, 197] But on the whole, the government and the contractors worked extremely well together on these programs, both in technical and in financial matters.

SUMMARY

Have Blue is a prime example of the value of an early, highly focused technology demonstration. According to Twigg, "I think the bottom line is that we were streamlined but very thorough, and perhaps most important, I had dedicated and expert people and had the full authority and charter to operate and succeed." [234] There were clear goals, and the program management was endowed with authority to achieve those goals. As a result, a breakthrough technology was convincingly demonstrated on a manned aircraft but without the large investment that accompanies a weapon system acquisition program. With the small initial investment, failure would not have precipitated a national scandal but success led to solid support at the highest levels for follow-on development of a revolutionary weapon system and substantially reduced the associated technical risks.

Chapter 3

ADVANCED TACTICAL AIRCRAFT GENESIS

By the spring of 1976, Have Blue phase I model test results had indicated a breakthrough of tremendous military value. There were still risks to be addressed, but an extremely high potential payoff had been identified. Accordingly, the Air Force initiated preliminary requirements studies of weapons systems based on the low observables breakthrough during the summer of 1976.[115, 188] By June 1977 a special project office had been established with strong support from Secretary of Defense Harold Brown and Director of Defense Research and Engineering (DDR&E) William J. Perry.[115] The overall goal of the U.S. Air Force low observables program was to develop the recently demonstrated breakthrough into a set of new capabilities for U.S. armed forces, as rapidly as possible and in total secrecy.

A hand-picked staff of five officers were briefed into Have Blue and "hidden" within AF/RDPJ, Room 5D156, the strategic reconnaissance office at Air Force Headquarters in the Pentagon. These officers were Col Dave Williams, Maj Ken Staten, Maj Joe Ralston, Maj Bob Swarts, and Maj Jerry Baber. Swarts and Baber were already assigned to RDPJ but had been working on other programs.[25] The new project office reported directly to the Deputy Chief of Staff for Research and Development (DCS/R&D) Lt Gen Alton D. Slay. Later, this was Lt Gen Thomas Stafford, and by 1979 the position was held by Lt Gen Kelly Burke. Maj Gen Robert Bond, who later became Vice

Commander of the Air Force Systems Command, also played a key role in the formulation of the low observables effort at this stage.

The project office operated under Special Access security, and a separate procurement function was set up to conduct contracting activities associated with low observables. Captains Robert Bird and John Hartford from the Air Force Office of Special Investigations (AFOSI) were brought in to structure the security aspects of the program.[234] Technical support came from the Air Force Systems Command's Aeronautical Systems Division (ASD) at Wright–Patterson Air Force Base, essentially the same cadre that was involved with Have Blue, but program management responsibility was primarily in RDPJ where all key actions originated during the formative stages of the program.[128]

The project office studied several potential applications of low observables technology, and a small number of specific programs were soon initiated. The second one of these was a manned strike aircraft.[25, 89] Maj Ken Staten became the acting program manager for the strike aircraft, which during its conceptual stages bore the very nonspecific label of Advanced Technology Aircraft (ATA). A probable reason for this is that even if the name were disclosed outside of Special Access channels, it would not reveal anything about the nature of the project.

TECHNOLOGY FOR A VERY LOW OBSERVABLE (VLO) WEAPON SYSTEM

While the Have Blue program was promising spectacular results, it did not address all of the technology needs for a VLO military aircraft. Have Blue focused on reducing signatures of the basic air vehicle and eventually proving that the aircraft could fly without detection. One of the tasks for the new project office, therefore, was to make sure the remaining technological pieces were in place to produce a VLO aircraft capable of performing a useful combat mission: locating, tracking, and launching weapons against a target without detection. Accordingly, in April 1977 Lt Gen Slay asked the Air Force Systems Command to identify applicable ongoing or planned technology efforts, noting: "While some success has been achieved in masking airframe and engine signatures, the enemy's increasing capability to detect and localize avionics emissions from strike/attack and air-to-air aircraft is of concern."[81]

Lt Gen Robert C. Mathis, vice commander of the Air Force Systems Command (AFSC/CV), responded,

> We have screened our current technology efforts that have the potential to reduce or eliminate detectable emissions from aircraft. Ongoing efforts which specifically address low signature avionics are quite limited. However, as a result of previous efforts, there are avionics stealth technology programs scheduled to begin in fiscal years 1977 and 1978. Existing and planned programs directly applicable to the low signature avionics problem are summarized in the attachment.[152]

The programs identified in this letter follow:

1) *Air-to-Surface Technology Evaluation and Integration* (18 month study, June 1976 to December 1977): Design studies and technology surveys are the intent, including concepts such as metallic radomes for reduced RCS and millimeter wave radars for reduced emissions.

2) *LPI Radar* (36 month study and demonstration, September 1976 to August 1979): The focus is on low probability of intercept radar, primarily for air-to-air applications.

3) *Bistatic Imaging Radar Studies* (12-month study, December 1976 to December 1977): The goal is a "cooperative air-to-ground weapon delivery" using illuminating vehicle (active) and attack vehicle (passive). Study will analyze imagery from bistatic data collection (see program 4, which follows).

4) *Bistatic Data Collection* (eight months, December 1976 to October 1977): The purpose is to compile bistatic measurement data (i.e., the receiver position is different from the transmitter) from test range results.

5) *FLIR [Forward Looking Infrared] Technology Demonstration* (24 month demonstration, April 1977 to March 1979): "The employment principle is to use a passive infrared sensor to eliminate emissions and thereby enhance survivability for air-to-ground fire control.... The significant constraints on this program are substantially improving FLIR resolution and sensitivity while restricting the available aperture.... Primary anticipated vulnerability will be from the FLIR lens system caused by off-board illumination."

6) *Avionics and the Stealth Vehicle* (six month study, October 1977 to March 1978): The purpose is to address the question of whether a vehicle can perform a useful military mission and still be stealthy.

7) *Stealthy Airborne Radar* (nine month study, July 1977 to March 1978): The focus is on air-to-ground radar.

8) *Laser Signal Reduction* (24 month demonstration, October 1977 to September 1979): The goal is to reduce a "precursor" signal emitted by existing laser designators, which can allow the designators to be "spoofed." Not specifically related to improving the survivability of the attacking vehicle, but rather to improve resistance to spoofing. The result will be integrated into a PAVE TACK laser designator pod for test and evaluation.

9) *Air-to-Air ARM Countermeasures* (study, results expected October 1977): The objective is to develop effective countermeasures against air-to-air antiradiation missiles.

In hindsight, we know that almost the entire avionics development effort on the F-117A focused on the development and integration of a passive infrared system for fire control, the subject of program 5. Relative to the preceding description, it is interesting to note that the FLIR turret on the F-117A can pivot so as to face backward when the FLIR is not in use, thereby "hiding" the lens system as well as minimizing the chances of damage to the lens components during flight.[49, 229] The other technologies on the list were too immature to be included in a weapons system at that time.

ATA Design Studies

During 1977 DARPA developed two sets of preliminary requirements for the ATA. One was for a "small" F-15 sized single-seat attack aircraft, intended to carry a 5000 lb payload (two nominal 2000 lb weapons plus suspension equipment) a distance of 400 n mile. The other was a two-seat bomber, closer in size to the FB-111, which would carry a 10,000 lb payload on a 1000 n mile mission.[34, 115] These were referred to as ATA "A" and ATA "B" respectively. A formal concept definition contract for $11.1 million was awarded to Lockheed Advanced Development Projects (ADP) on October 10, 1977, for a one-year study based on the two sets of requirements.[128] Detailed operational concepts, technical requirements, cost, and schedule estimates evolved during this study and two preliminary designs were produced. Properties of the two designs, as of February 1978, are summarized in Table 2.

Wind tunnel tests of the candidate ATA designs were conducted as part of the concept exploration studies; configuration wind tunnel tests were initiated in early 1978. Test results from NASA Ames were available by August. Inlet testing also was conducted during 1978 with a one-fourth-scale inlet test completed in June or July.[232]

As seen in Table 2, by early 1978 the payload for the "B" model was reduced to 7500 lb. Early sizing studies had indicated that the "B" model would need six engines, not just four, to achieve its original range–payload goals.

Table 2 Characteristics of ATA "A" and ATA "B"[125]

	ATA "A"	ATA "B"
Mission radius	400 n mile	1000 n mile
Payload	5000 lb	7500 lb
TOGW	43,000 lb	90,000 lb
Length	64 ft	76 ft
Wingspan	43 ft	47 ft
Crew	1	2
Number of engines	2	2
Types of engines	GE F404 (nonafterburning variant)	GE F101 ("partially" afterburning)
Costs	Projected by Lockhead to be about 50% higher for the "B" aircraft than for the "A," both in FSD and in production.	
General configuration	Similar to Have Blue except for butterfly tail (tails mounted near the centerline and slanted outward instead of inward as on Have Blue).	

This is because an aircraft's gross weight grows roughly exponentially as a function of its design range. As the design range increases, so does the rate of weight growth. It simply would not be practical to build a large, long-range aircraft with the low lift-to-drag ratio of Have Blue. Although a higher lift-to-drag ratio could have been achieved through various design changes, some Air Force officials and even some designers and managers at Lockheed believed that the Have Blue breakthrough was a result of "black magic" rather than science. These individuals strongly resisted any significant departures from the basic Have Blue configuration. As a result, it was necessary to reduce the payload of the "B" model to keep the air plane's size somewhat reasonable.[34]

Program Direction

Many of the decisions that characterized the F-117A program were made during this period. A Lockheed budget and schedule proposal summary prepared in February 1978 included the following information.[125]

1) Research, development, testing, and evaluation: Five aircraft, two to be subsequently upgraded to operational status.

2) Production: 20 initially (funded in two lots of 10 each in FY80 and FY81) with a possible 100 additional follow-on aircraft.

3) Early initial operational capability was a high priority; desired date was early 1982.

4) Production go-ahead prior to flight test completion; level of concurrency similar to that on the successfully accomplished SR-71 program.

5) Logistics concept "remove and replace," with repairs at the factory; again similar to the logistics concept used in the SR-71 program.

6) Use of existing engines with minor modification. These would be General Electric F404s for the "A" aircraft or General Electric F101s for the "B" aircraft. The plan, if ATA "B" were pursued, was to "cannibalize" engines for research, development, testing, and evaluation from the recently canceled B-1A program.

7) Maximum use of off-the-shelf subsystems and components. "We have assumed that one-third of the avionics will be development and two-thirds off-the-shelf or used with minor modification."

The cover letter on this document from Lockheed ADP General Manager Ben R. Rich to Lt Col Ken Staten stated that,

> As requested, we have prepared data for two alternative program schedules. The first schedule is not constrained by government funding limitations and results in an initial operational capability in January 1982. The second schedule is based on a...limit to first year funding. Under this schedule initial operational capability is one year later. I believe that the most viable program may be based on a schedule that lies somewhere between these two alternatives.[125]

In June 1977 President Carter had canceled the B-1A program. Filling the B-1's role became a critical objective for the Air Force during the next several months, and this focused attention on the "B" model ATA. Although this larger version became the strong favorite with the Air Force and particularly with the Strategic Air Command (SAC), Lt Col Joe Ralston and Lt Col Ken Staten at RDPJ voiced the realistic, but in some quarters unpopular, view that the "B" model stealth bomber was not technically feasible at that time.[115]

This opinion was based on the risks inherent in scaling the Have Blue concept up to a size as large as the ATA "B." As noted earlier, even though the requirements had been relaxed, the designers knew that the "B" was right at the upper bound of what was realistically achievable.*

After the payload weight had been reduced to 7500 lb, another difficulty involved insufficient justification for the added cost and risk of the "B." Ralston and Staten recommended deferring the bomber and focusing on the smaller attack aircraft. SAC headquarters did not like this, but Gen David C. Jones, who had become chairman of the Joint Chiefs of Staff, and Gen Lew Allen, who had succeeded Jones as Air Force Chief of Staff, backed up Ralston and Staten on the basis that valuable information might be gained from pursuing the lower-risk, less expensive "A" model before attempting to build a larger bomber.[115] This was also the path that would introduce a low observable weapon system into the operational inventory as rapidly as possible, consistent with the overall goals of the U.S. Air Force low observables program. In the summer of 1978 Air Force officials, therefore, decided to discontinue the "B" and to take the "A" model Advanced Technology Aircraft into full-scale development.†

---

* Their conviction was reinforced during a visit to Lockheed's Rye Canyon research facility. Col Williams, Lt Col Ralston, and Lt Col Staten each made five takeoff attempts with the ATA "B" on Lockheed's flight simulator. According to Lt Col Jerry Baber, "Ralston saved it once; all other takeoff attempts resulted in the loss of the [simulated] aircraft."[25] All three men were experienced fighter pilots.

†The Air Force and Lockheed continued to study a bomber under the new heading of Advanced Technology Bomber (ATB). An unsolicited proposal was also received from Northrop. Initially, Lockheed was considered the lead, with Northrop as a fallback, but by early 1981 the effort had become a full competition.[92] During the years since Northrop had been eliminated from the XST program, the company had made significant advances through the DARPA-sponsored Tacit Blue program, which was nearing the flight test phase at that time.[5,29,63] The Northrop ATB design utilized a relatively high-aspect ratio wing with efficient airfoils, while Lockheed was still proposing a faceted ATB design crudely resembling a scaled-up F-117. Meanwhile, William Perry (DDR&E) and Col Paul Kaminski (special assistant to DDR&E) had convened a Defense Science Board task force to look at future investments in low observables. The ATB source selection was in progress when the DSB task force issued its report, which suggested that the ATB requirements were too narrow and in particular did not call for any low altitude capability. While the Lockheed design met the initially stated requirements, Northrop's proposal was considered to exceed them and was more in keeping with the recommendations of the DSB task force to pursue broader capabilities. Northrop was therefore selected later in 1981 to build what became the B-2.[115]

Chapter 4

Full-Scale Development

Initiation

Once the Air Force decided to proceed with full-scale development, funding requirements for the program increased substantially. The government established covert funding channels to support the Advanced Tactical Aircraft (ATA) and other low observable systems that were moving beyond concept exploration and into development. Key congressmen and staff members on the four applicable oversight committees—the House Armed Services Committee (HASC), the Senate Armed Services Committee (SASC), the House Appropriations Committee (HAC), and the Senate Appropriations Committee (SAC)—were briefed into the program. The staff at RDPJ (the Air Force low observables office in the Pentagon) began to process the funding requirements and necessary documentation. This included determination and findings and sole source justifications to authorize the award of the full-scale development contract to Lockheed's Skunk Works without a full and open competition on the basis of the engineering work already performed by the

company during the Have Blue program and the recent ATA design studies. The program then was given the name "Senior Trend." By November 1978 the necessary preparations had been made and Lockheed was awarded the full-scale development contract.[25, 115]

The cost-plus-fixed-fee contract, initially negotiated at $340 million, covered five completed aircraft plus spares, support, and flight testing. The initial cost estimate for the total program, including government test support costs, program office costs, and government furnished equipment including the engines was $505 million in FY79 dollars.[90]

Senior Trend program objectives were to develop an attack aircraft to be used against high-value targets with very low observability as its primary design goal, field the system as rapidly as possible, and conduct the entire program in total secrecy.[87]

Officials at the highest levels of the U.S. government formally decided that the Senior Trend program and the low observables effort as a whole would be executed in a streamlined and covert manner outside of the normal acquisition process.[216] This involved implementing Special Management procedures. However, Special Management arrangements did not allow statutory requirements or executive orders to be violated; formal Air Force guidelines laid out Special Management procedures to be followed including what procedures and documentation could be bypassed. At the time Senior Trend was initiated, these guidelines were apparently contained in Air Force Regulation AFR 66-22, which was superceded by AFR 800-29 in 1982.[238] Special management was characterized by face-to-face communication among the various government agencies and the contractor, small staffs, abbreviated oversight, and short direct lines of authority. Access to program acquisition documentation was strictly controlled. Work was not delayed by bureaucratic and administrative details because of a new streamlined procurement and contracting authority.

> This enabled both the government and the contractor to make rapid decisions and to operate with small staffs and simplified paperwork systems. For example, the initial full-scale development contract consisted of only 70 pages including the statement of work. Lack of paperwork was compensated for by close and rapid communications on a daily basis with personal contacts and a secure electronic message capability.... Representatives of the Air Force project office visited the Skunk Works no less than weekly and had essentially full-time residency as the program reached its critical stages.[128]

The highly streamlined approach led to extremely effective program management.

The project proceeded with a high degree of concurrency to achieve the earliest possible capability. The original target initial operational capability date was March 1982, or 40 months from the start of full-scale development.[199] The prime contractor, Lockheed, had experience with special programs involving compressed schedules, streamlined management procedures, and tight security. The SR-71 program was a recent example, and according to Lockheed, "The fact that Lockheed Skunk Works had successfully executed that program gave credibility to our proposals and was a factor in the Air Force decision to proceed with the [Senior Trend] program." [128] The ongoing risk reduction accomplishments of Have Blue, which had begun to provide in-flight signature measurement data by the initiation of Senior Trend's full-scale development, provided additional confidence in Lockheed. Senior Air Force and Office of the Secretary of Defense staff were continually informed of the Have Blue program and its accomplishments.

Requirements

Senior Trend's performance requirements were not at all ambitious relative to conventional attack aircraft of the time except in the area of signatures. Furthermore, requirements allowed for flexibility and performance tradeoffs as long as the essential mission capabilities included specified radar cross section at specified frequencies, polarization and aspects, weapon delivery accuracy, and a mission radius of 400 n mile with a 5000 lb payload (normally consisting of two 2000 lb bombs). Other less critical goals included takeoff and landing performance, a 6 *g* limit load factor, and a maximum speed of 0.9 Mach.[179] The radar cross section (RCS) requirements were based on the proven performance of the Have Blue demonstrators:

> With Have Blue in-flight RCS data, 6/10 scale model data from RATSCAT, and ECHO computer program data [ECHO was Lockheed ADP's RCS prediction program], a "soft" spec was built as a general guideline for the F-117A.[183]

The preceding statement should be qualified by noting that the Have Blue program was based on signature goals derived from engagement simulations and analysis of the various threats conducted by the DARPA study contractors and by the Air Force (RPV SPO) during the period leading up to Have Blue.[186, 234] Several earlier studies sponsored by various agencies had been conducted on a "how-good-can-we-do" basis and had led nowhere.[257] What set Have Blue apart from those earlier efforts was that the program focused on achieving signature goals that really would provide high survivability against most known threat systems. Contractors who merely said "This is the best we can do" were quickly eliminated. The consequent breakthrough enabled the Have Blue program to achieve the necessary signature levels. Once this was accomplished, the Senior Trend program was not required to push low observables performance any farther but only to achieve similar RCS values in a militarily useful aircraft. This would "provide the National Command Authority with a capability to strike high-value targets at night in various threat environments...."[64, 78]

Program Office

The low observables programs, including Senior Trend, that were progressing from concept exploration into development transferred from AF/RDPJ in the Pentagon to the Air Force Systems Command's Aeronautical Systems Division (ASD) at Wright–Patterson Air Force Base, Ohio. Personnel in the newly created System Program Office (SPO) reported directly to the ASD commander. The office was initially called the Classified Aeronautical Systems Program Office, then after December 1981 the Directorate of Low Observables.[35] The entire effort operated under special management guidelines. Maj Peter Knauth was assigned as security manager at the new SPO, and he is credited with making a significant contribution to the excellent security that was maintained throughout Senior Trend and the other low observables programs.[234]

Col David B. Englund became the first director of the new SPO. Prior to this assignment, Col Englund had been program manager for the Advanced Medium STOL Transport (AMST), a technology demonstrator program that had been conducted under a DoD-wide "experimental prototyping" initiative of the early 1970s, and resulted in developing and flight testing of the YC-14 and YC-15 STOL demonstrator aircraft. The AMST and other programs conducted under that initiative (including the Lightweight Fighter program) embodied many of the desired characteristics for the low observables effort: direct, flexible management; abbreviated oversight; face-to-face communication; real-time decision making; and emphasis on demonstration rather than documentation. Col Englund helped establish the streamlined procedures that the new SPO would adopt.[25] Both the experimental prototyping initiative and the early U.S. Air Force low observables effort initiated a variety of projects with an understanding up-front that only the most promising ones would move into production. This fostered an environment of risk management rather than risk avoidance.

The Classified Aeronautical Systems Program Office structure was typical for an office that managed multiple programs, and it was quite small. It included an overall front office; then each program had its own "front office" with a dedicated staff of about six. All programs shared a support staff of four to six people per discipline (including contracting, finance, administrative, security, plus the various engineering disciplines). The individual program managers had decision authority for their programs and reported to Col Englund for administrative purposes only. The organization of the SPO as of the summer of 1984 (by which time it was called the Directorate of Low Observables) is illustrated in Fig. 15. The numbers in parentheses refer to the number of personnel dedicated to each program or area.[35]

At this time the Advanced Technology Bomber (ATB, later to become the B-2) had just split off into a separate office. The remaining programs were Senior Trend, the Tri-Service Standoff Attack Missile (TSSAM), Tacit Blue, and the Low Observables Technology Base. The latter was initiated by then-Col Paul Kaminski when he became head of the RDPJ office at the Pentagon in 1981 to ensure that as specific programs moved into full-scale development and production they would not impede advancement of basic low observables technology.[110, 115] Although the details of the specific programs and the exact numbers of people in each area changed with time, the overall structure shown in Fig. 15 is fairly representative of the Classified Aeronautical Systems Program Office in the early stages of Senior Trend.

In spite of the small size of the engineering support staff, considerable technical expertise at ASD was available to draw upon as needed:

> All ASD engineers were assigned to the home office of ASD's Directorate of Engineering (ASD/EN). When program offices needed engi-

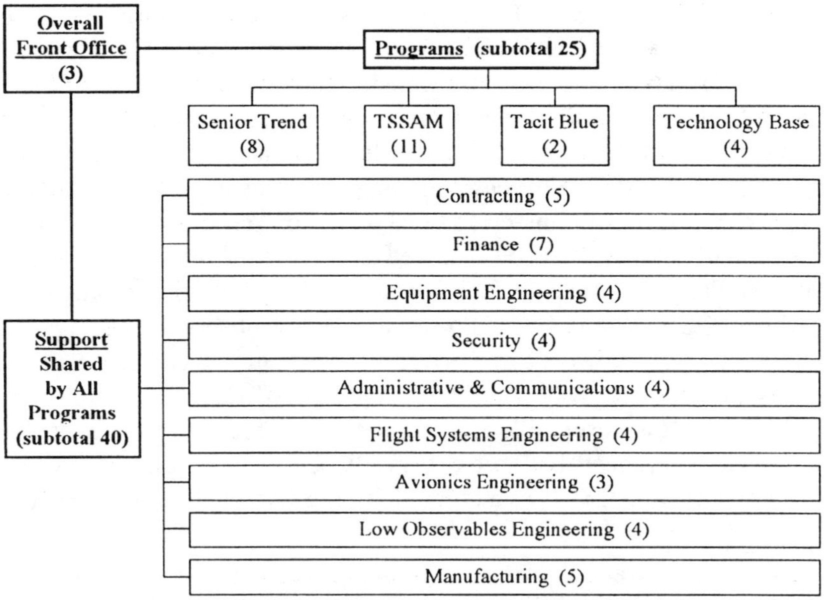

Fig. 15 Directorate of Low Observables, summer 1984.[35]

> neering expertise in a given area, the director of EN, Fred Rall, would review their requirement and select the best candidate who would then be matrixed to the requesting office.
>
> When stealth program managers needed an engineer with specific skills they followed much the same process.... A LO [low observables] program manager would describe the skills he needed, or, more often, would tell Rall exactly who he wanted. Although Fred Rall was not briefed into the stealth programs he realized that there were offices within ASD that were doing exceedingly important work that required the highest calibre people and he would grant the request—no questions asked. Although the engineers were still assigned to ASD's EN office, they were given complete autonomy after being matrixed—and disappearing—into stealth programs.[40]

Col Eldred D. (Don) Merkl was the first Senior Trend program manager. Merkl had a doctorate in engineering, having written his dissertation on the flight control system of the F-16.[34] He had taught control theory at the Air Force Academy and the Air Force Test Pilot School. All of this was valuable experience because flight control was one of the key challenges in the program, and in fact the F-16 flight control system was adapted for both Have Blue and the F-117A. Merkl served as program manager until August 1981 when he retired from the Air Force and was replaced by then-Col Richard M. Scofield.* The Senior Trend program managers from 1978 through early 1990 are shown in Table 3.

In June 1979 all Senior Trend contracts previously administered out of RDPJ were transferred to Wright–Patterson Air Force Base. Contracting Officer Lt Col Jerry Baber was also transferred to the Senior Trend SPO.[25] As the SPO built up and the Have Blue program tapered off, "the members involved in the phase down of Have Blue and the start-up of ATA/Senior Trend were collocated. Key engineering personnel with responsibilities in low observables...were transferred to ATA/Senior Trend. The same comment applies to the highly dedicated Lockheed team," according to former Air Force Have Blue Program Manager Jack Twigg.[234] Along with the technical expertise, lessons learned from Have Blue regarding low observables design, maintainability, and testing were carried over into the Senior Trend program. Twigg stayed with this office following the completion of Have Blue. He managed Tacit Blue, the low observable aircraft program. This Northrop-built aircraft demonstrated advanced battlefield surveillance technologies in a very low observable airframe. Later, as a colonel, Twigg was the low observables sys-

*In June 1983 Scofield became program manager for the Advanced Technology Bomber (ATB, which became the B-2). The ATB split off into a separate office in the summer of 1984.[35] Later, as a lieutenant general, Scofield was the commander of the Air Force Systems Command's Aeronautical Systems Center (ASC, the organization which was formerly ASD), prior to his retirement in June 1996.

Table 3 Senior Trend program managers, 1978–1990[35, 243]

Name	From	To
Col Eldred D. (Don) Merkl	Dec. 1979	Aug. 1981
Col Richard M. Scofield	Aug. 1981	June 1983
Col John J. (Jack) Colligan	June 1983	July 1985
Col Vincent T. Baker	July 1985	June 1987
Col Berwyn A. Reiter	June 1987	Aug. 1990

tem program director and thus head of the entire ASD low observables office until he retired in 1983.[234]

The RDPJ office at the Pentagon continued to plan and budget activities as well as interface with Congress, the Office of the Secretary of Defense, and Air Force headquarters. Col Paul Kaminski* was assigned to RDPJ in March 1981, formerly having been special assistant to William J. Perry, director of defense research and engineering.[115] The relationship between RDPJ in the Pentagon and the SPO at Wright–Patterson Air Force Base was such that RDPJ made every possible effort to answer planning, budgeting, and other administrative issues and inquiries internally rather than passing on tasks to the SPO. "Those of us in Washington, D.C., didn't pass 'what ifs' to the SPO unless we were 90% sure they would happen," according to Col Kaminski. The RDPJ staff was "small but very well connected to the system" and, therefore, well equipped to handle program coordination in the Washington area. "The SPO did not have to waste time coming to Washington, D.C., to give briefings."[116] This left the SPO free to concentrate on the technical management of the program with a minimum number of distractions.

The SPO maintained an acquisition plan with a brief description of the system, program history, priorities, funding profile, delivery requirements, summary of schedule and technical risks, contract types, acquisition milestones, logistic support concept, and a list of participants in the plan's preparation (usually the program manager, business manager, budget analyst, contracting officer, and contract negotiator).[240–243] The plan was revised as necessitated by major changes in the program such as acquisition milestones or changes in the number of aircraft to be procured. Program priorities listed were security, observables, schedule, other system performance, and cost. The order varied, but security was always first, followed by observables, and then the other three. Attached to the acquisition plan would be a U.S. Air Force Headquarters' Determination and Findings document, one each for

* At time of writing, the Honorable Dr. Paul G. Kaminski is Under Secretary of Defense for Acquisition and Technology (USD/A&T).

airframes and engines, justifying the procurement of those items on a sole-source basis from Lockheed and General Electric respectively: "AFSC [Air Force Systems Command] proposes to acquire... 20 ATAs and related initial spares and repair parts, aerospace support equipment, contractor support, training, and data at a total estimated price of...without formal advertising under 10 U.S.C. 2304(a)(14)." [113] The wording of these one-page Determination and Findings changed little from one version of the plan to the next except for dates, prices, and quantities. An early example of a Senior Trend acquisition (Plan No. RWSX-80-3, dated July 31, 1980) is presented in its entirety in Appendix E.

AIRCRAFT DESIGN

The F-117A team at Lockheed ADP included principal engineers Alan Brown (chief engineer, low observables, and program manager), Dick Cantrell (aerodynamics), Bob Loschke (flight controls), Ed Baldwin and Henry Combs (structures), and Manufacturing Director Bob Murphy, all of whom were veterans from the Have Blue program.[175, 181, 232] Subsequent Lockheed program managers for the F-117A are shown in Table 4.[181, 232]

The F-117A is illustrated in Fig. 16, together with the Have Blue aircraft for comparison. As noted earlier, Have Blue was not a prototype, but it had validated the overall concept and had proven the ability to achieve the desired signature levels in conjunction with acceptable flying qualities. Summary characteristics of both aircraft are shown in Table 5.

The F-117A was large for a single-seat aircraft, roughly the same size as the McDonnell Douglas F-15. Apart from its size, the most obvious external difference between the F-117A and the Have Blue demonstrator was in the design of the tail surfaces, which on the F-117A were mounted at the centerline and canted outward rather than mounted outboard and canted in. The primary reasons for this change were to improve directional stability (by locating the

Table 4 Lockheed program managers for the F-117A

Name	From	To
Alan Brown	Nov. 1978	Jan. 1981
Norman Nelson (earlier Have Blue program manager)	Jan. 1981	Feb. 1982
Sherman Mullin	Feb. 1982	Oct. 1985
John Sheridan	Oct. 1985	April 1988
Gary Riley	April 1988	June 1989
Paul Martin	June 1989	April 1996
Ross Reynolds	April 1996	Present

Table 5 F-117A and Have Blue characteristics (from Ref. 104, except as noted)

	F-117A	Have Blue
Length	66'	38' (w/o noseboom)
Wingspan	43'	22'6"
Height	12'6"	7'6"
Wing area	1140 sq ft	386 sq ft
Wing sweep	67.5 deg	72.5 deg
Empty weight	29,220 lb	8950 lb
Maximum fuel weight	18,280 lb	3500 lb
Payload	5000 lb (2 weapons)	none
Maximum takeoff weight	52,500 lb	12,500 lb
Range/endurance	400 n mile unrefueled mission radius[179]	1 h
Maximum speed	0.9 Mach (limited to 0.8 initially)	0.8 Mach
Propulsion	2 × General Electric F404-GE-F1D2 10,800 lb thrust each	2 × General Electric J85-GE-4A 2850 lb thrust each

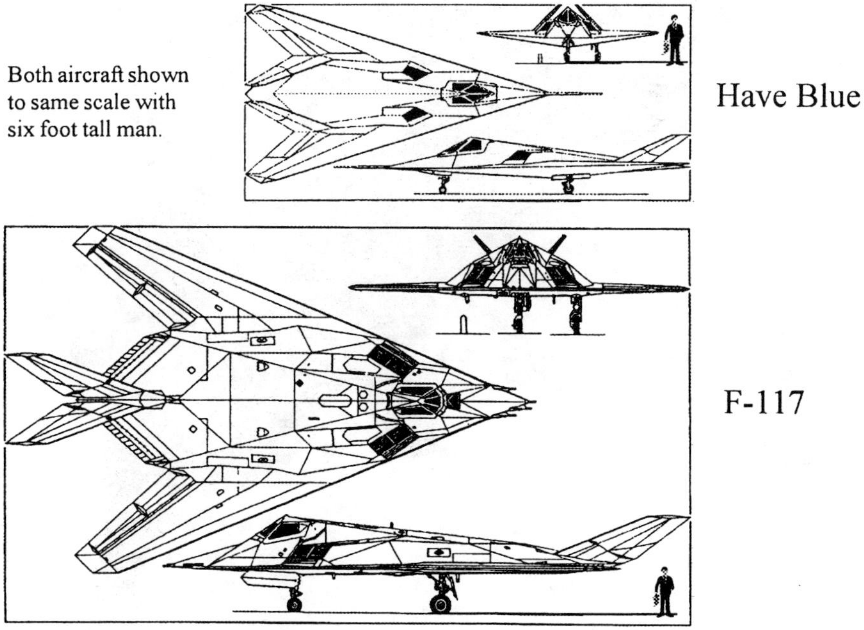

Fig. 16 F-117A three-view, with Have Blue for comparison.[104]

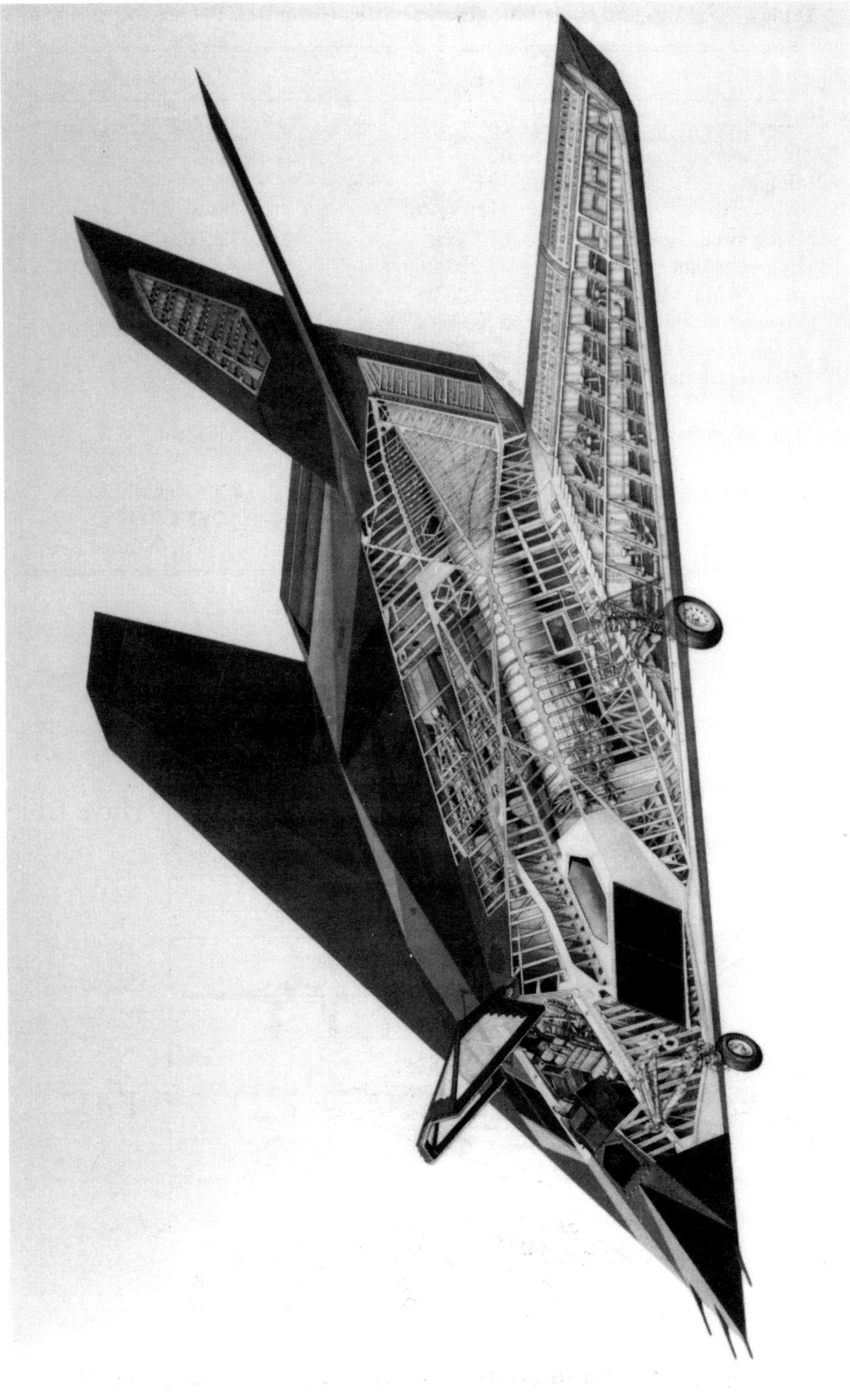

tailfins further aft, thereby achieving a longer moment arm about the center of gravity); to improve the structural design by mounting the fins on the center keel of the aircraft; and to alleviate problems associated with heating of the lower surfaces of the fins, which in the original configuration were in close proximity to the upper deck of the platypus. The term "platypus" was used variously to refer to the wide, flat exhaust nozzle installation used on the Have Blue and F-117A aircraft or specifically to the part of the lower lip of the nozzle that extended behind the nozzle exit to conceal it from view.[34, 212]

The ATA design included the outward-canted fins at least as early as February 1978 (see Table 2 in Ref. 125), which was less than three months after the first flight of Have Blue No. 1 and several months before the first flight of Have Blue No. 2. The need for the change must, therefore, have been determined through analysis and ground testing or have been almost immediately apparent from the first few Have Blue flights. Other significant changes from the Have Blue configuration were a reduction in wing sweep (although the F-117A's wing sweep of 67.5 deg is still extremely high for a subsonic aircraft), provisions for weapons bays in the central fuselage region and sensors in the nose, and a fixed platypus trailing edge rather than the movable platypus control surface used on the Have Blue aircraft. While eliminating the platypus control surface, the F-117A design added outboard elevons on the wings to provide sufficient pitch control authority. Have Blue had only used inboard elevons.

ENGINES

The engines for the F-117A were derived from the General Electric (GE) F404 turbofan used in the McDonnell Douglas F/A-18 naval fighter. The variant for the F-117A was designated F404-GE-F1D2. The chief modification was the deletion of the afterburner. Engines were treated as government furnished equipment (GFE) and procured under a separate CPFF contract with GE. It was initially planned that 13 engines would be procured for the full-scale development phase and 60 for production (for the initially planned 20 production F-117As). The contract between the government and GE for the F404-GE-F1D2 engines was unclassified, and there was no special management carve-out for this contract. However, ground rules were set during contract negotiations in late 1978. Lockheed and ADP would not be mentioned in any unclassified correspondence. The identity of the program office would not be mentioned and interface data must be innocuous. The schedule must remain open, in other words GE required the freedom to deliver engines as they were completed, because there was no place to "hide" the engines at GE. Finally, any correspondence between GE and ADP would be hand-carried by the GE representative in Los Angeles.

The cover story at GE would be that the engines were for a "classified re-engineering program."[66,67] R. K. Bertelson became the GE program manager for the F1D2 engine. Reportedly, only Bertelson and two other, higher-ranking senior executives at GE were read into the Senior Trend program and knew what the engines were really for.[34]

Subsystems

To speed development, the F-117A, like Have Blue, used off-the-shelf subsystems and components wherever possible. As already noted, the engines were a nonafterburning version of the GE F404 engine used in the McDonnell Douglas F/A-18. The environmental control system (ECS) was from the Lockheed C-130, brakes were from the Grumman Gulfstream II business jet, and other landing gear components were from the F-15. Cockpit displays were from the McDonnell Douglas F/A-18, and other cockpit components were taken from various Lockheed aircraft. The electrostatically gimballed Honeywell SPN-GEANS inertial navigation system (INS) from the Boeing B-52 was used for navigation. This had caused some concern within the Air Force due to a perception that the SPN/GEAN system might prove unreliable and difficult to maintain. However, according to Col Ted Bowlds, who as an Air Force captain was assigned to the F-117A Joint Task Force:

> ...Contrary to what we had been led to believe, the SPN/GEAN navigation system proved to be very reliable. System testing would later consist of several upfront dedicated navigation flights followed by periodic gathering of data on subsequent flights. The accuracy of the system matched very closely the mission profile of the F-117A (i.e., in about an hour's worth of flight, the drift was acceptable).[31]

Targeting System

The infrared acquisition and designation system (IRADS) was designed to provide a capability to detect, track, and designate targets without generating

any detectable emissions from the aircraft. The IRADS was based on two forward-looking infrared (FLIR) targeting units from the F/A-18. Each of these units included an infrared sensor and a laser designator, mounted in a swiveling turret. In a conventional application, the turret was mounted on the front end of an external pod, providing it with a large range of motion. On the F-117A, turrets needed to be placed inside the aircraft behind screens that were specially designed to block radar frequency radiation while allowing the higher-frequency infrared and laser radiation to pass through. The internal mounting arrangement restricted the available field of view, so two units became necessary, one looking primarily forward and one looking primarily downward, as shown in Fig. 17.

After using the inertial navigation system to navigate to the target area, the pilot acquired the target with the FLIR unit and designated it; the target was then automatically tracked and illuminated by the laser designator. As the aircraft moved closer to the target, the IRADS automatically switched from the FLIR to the downward-looking infrared (DLIR) to ensure continuous

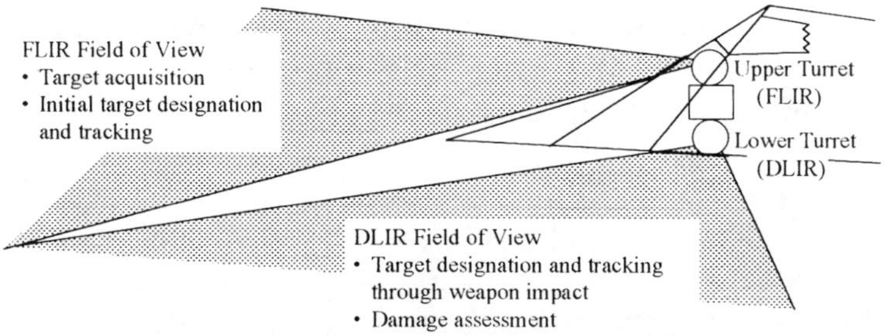

Fig. 17 Conceptual diagram of IRADS system operation (not to scale).

target tracking. Then, as the weapon, normally a laser-guided bomb (LGB), was released and the aircraft flew over the target, the system used the DLIR unit to keep the laser spot on the target up until the moment of weapon impact. Video imagery of the entire attack profile was recorded and provided valuable input into post-strike bomb damage assessment.

Vehicle Shaping/RCS Prediction

As noted, the F-117A used the same basic concepts in shaping as the earlier Have Blue aircraft. However, it was not simply a scaled-up Have Blue; it was a new design. The ECHO computer program was relied on heavily to develop this design with confidence that it would meet its RCS requirements. The program had been enhanced in several ways since Have Blue. In the early stages of Have Blue, the designers were limited on the number of facets that ECHO could handle.[204] By the time the F-117A was designed, the limits had been decreased so that this was no longer an issue. In addition, two important theoretical improvements had been incorporated into ECHO: edge diffraction calculations, and a limited ability to calculate the effects of nonmetallic materials.

The original version of ECHO, used to design Have Blue, approximated surface currents and did not account for presence of edges. It was shortly *after* the initial design of Have Blue that Denys Overholser, the lead developer of ECHO, learned of the Physical Theory of Diffraction developed by P. I. Ufimtsev in the Soviet Union during the 1950s and early 1960s.[186] Although Ufimtsev's results are not directly applicable to calculating diffraction for edges at arbitrary aspects,[123, 176, 186, 237] the theory did provide valuable conceptual and mathematical techniques for dealing with the electrical currents near the edges of a panel, which allowed Overholser and his team to eventually incorporate diffraction calculations into ECHO.[186]

During the same period, ECHO was also improved to account, at least empirically, for some effects of nonmetallic materials. Initially, ECHO had predicted the signature of a shape *assuming a perfect metal skin*. The effects of radar absorbent coatings and special edge constructions had to be considered separately. Improvements allowed ECHO to incorporate reduction factors to the reflections from coated surfaces. The reduction factors were determined experimentally, not analytically: each different coating material was assigned a reduction factor based on measurements of material samples.[34]

Edges presented a more difficult problem. To build an experimental database covering all possible edge configurations and orientations was not practical. However, as noted in Chapter 2, special edge materials and construction techniques used by Lockheed had been observed to reduce the edge scattering to approximately the values predicted by the basic physical optics theory (ECHO I) without the additional diffraction contribution. During the design of the F-117A, the basic ECHO I results, therefore, were used as a "best case" calculation for edges, and the full diffraction calculation was used as a "worst case." A prediction then was based on an "educated guess" somewhere between these extremes. It was not until the mid-1980s that Lockheed developed the capability to predict the performance of special edges analytically using the Method of Moments.[34]

Although the developments in RCS prediction capability were important, the role of testing must not be overlooked. Subscale configuration RCS models, as well as large-scale or full-scale RCS models of specific components (such as the engine inlets) were tested extensively during the design of the F-117A, both at Lockheed and at government test range facilities. The program had full support from RATSCAT whenever required, which was a significant improvement from the early XST days.[186] The first large-scale RCS test, of a 6/10 scale model, was conducted at RATSCAT in April 1979, approximately six months after the full-scale development go-ahead.[244] For reference, the Have Blue flight tests were nearing completion at that time.

Aerodynamics/Stability and Control

Wind tunnel testing in support of the F-117A design (then known as the ATA "A" had begun in early 1978.[232] Like Have Blue, the F-117A was designed to be unstable in pitch (relaxed longitudinal stability) and relied on a Lear–Siegler fly-by-wire flight control system modified from the General Dynamics F-16. Actuators, computers, and power supply were stock or only slightly modified. However, like Have Blue, the F-117A used new control laws, new air data interfaces and a different failure detection and redundancy management scheme than the F-16 to suit its particular arrangement of control surfaces and air data sensors, as well as unique aerodynamic characteristics. The F-117A also used a conventional control column, because a fixed, force-sensitive side stick controller, as used on the YF-16, had not been found satisfactory on Have Blue.[175] The production F-16 also abandoned the fixed side stick in favor of a side stick with some motion.[13]

The F-117A has six control surfaces: the two canted rudders providing yaw control, plus two elevons on each wing to perform pitch and roll functions. Although the F-117A was only designed to be unstable in pitch, the automatic control system operated in all axes. The flight control computers commanded the movements of all control surfaces by comparing the pilot's commands with what the aircraft was actually doing at any given time, as determined by the air data and inertial sensors. Air data was provided by four special faceted low observable pitot probes mounted at the nose of the air-

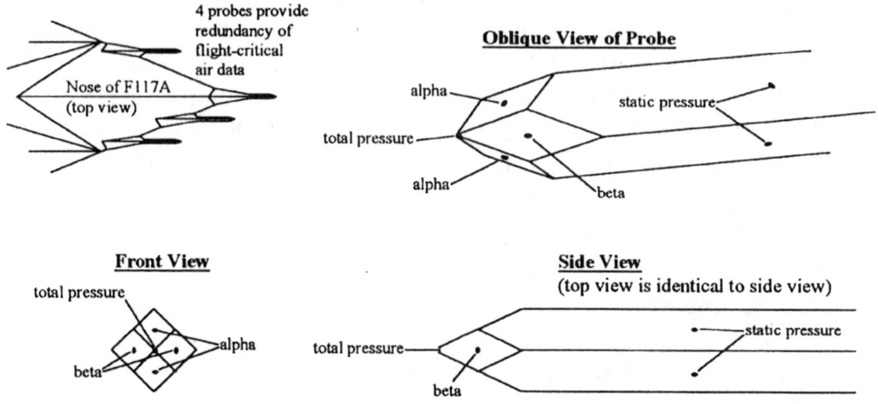

Fig. 18 Low observable air data probes on F-117A (not to scale).

craft, as illustrated in Fig. 18. Total pressure was measured from the single opening at the tip of each probe. The differential pressure between the upper and lower forward-facing ports was used to determine angle of attack α, while the differential between the right and left forward-facing ports is used to determine angle of sideslip β. Static pressure measurements are taken from each of the four ports located farther back along the sides of the probe, and averaged by the air data computer.[34]

The faceted shape and unusual layout of the Have Blue and F-117A were commonly viewed as making them unstable. As noted earlier in the description of Have Blue, a faceted, very low observable aircraft could have been designed with positive stability about all axes. However, faceting and other low observable shaping factors, such as very high wing sweep, make the flying qualities problem more challenging and result in some aerodynamic penalties, such as higher drag and reduced maximum lift. Lockheed designers took advantage of the emerging fly-by-wire technology to gain greater freedom in designing to the constraints imposed by signature considerations and still achieve acceptable flying qualities while minimizing the performance penalties. Thus, fly-by-wire control was desirable but not absolutely essential to the design of a faceted very low observable strike aircraft. However, fly-by-wire is not a cure-all, and cannot be relied upon to compensate for faulty aircraft design.

STRUCTURES AND MATERIALS

The F-117A had a conventional basic structure primarily of aluminum, except for the aft fuselage, which was constructed from titanium. Nonstructural radar absorbent material (RAM) was applied over the metal skin, except the tailfins. They were originally aluminum but later were replaced with a new thermoplastic design. The F-117A empty weight of approximately 29,000 lb

initially included about 2000 lb of RAM, although it is possible that lighter weight coatings have since been developed.

DEVELOPMENT AND FABRICATION

Construction of a wooden mockup started on January 1, 1979, and ended in early December.[232] Structures fabrication for the first flight test aircraft began in December 1979.[244] Also in late 1979, logistics activities and long-lead ordering began for the first production lot, and an operational unit—designated the 4450th Tactical Group—was formed.[49, 224, 240] During 1980 fabrication began on the static test airframe, referred to as Aircraft 779 (tail numbers of the flightworthy aircraft began at 780). A fatigue test airframe, which became Aircraft 777, was added later in the program.[232]

Almost immediately, materials and manufacturing difficulties slowed down progress. Have Blue experience had revealed just how much the RCS could be increased by the slightest gaps or violations of the vehicle's contours. Much tighter tolerances were therefore required for the F-117A than for a conventional airplane, both for the fixed features and edges of the airframe and especially for doors and access panels. Detail design and appropriate treatment of every external component of the aircraft was critical to minimize its contribution to overall aircraft RCS. Radio communication antennas had to be retractable to ensure stealth was maintained in high-threat areas. All openings in the aircraft's carefully tailored shape (for example the canopy, inlets, and sensor apertures) needed to be treated to behave electromagnetically as if they were part of the vehicle's surface. All of this required a highly disciplined approach. In some cases, achieving a low observable design was relatively easy, but subsequent ground or flight testing revealed problems with a component's durability or its ability to function under realistic flight conditions. Many of the challenges were predictable, but nevertheless it took time to overcome them. The original schedule was simply too optimistic.

Other unpredictable and often seemingly irrational problems occurred. For example, on one occasion a certain ferrite paint that Lockheed was using reportedly lost its radar-absorbent properties. The cause was that the supplier of the paint had made what they thought was a harmless change in their manufacturing process.[197] Although the Skunk Works carefully specified every part and product that vendors produced for them, the vendors often did not know for whom they were working, or for what their products were to be used.[216] Because low observables was a relatively new and little understood field, Lockheed found some difficulty in accurately specifying to the vendors exactly what was needed, thus causing a lot of "cut-and-try" in the early stages of development.[235]

One of the most challenging areas was the tailpipe. Most tailpipes are circular in cross section, which is structurally the most efficient shape for containing the high pressure exhaust gases. The tailpipe of the F-117A transitions from a circular cross section at the engine to a wide, flat cross section at the nozzle as shown in Fig. 19.

The flattened cross section experiences greater stresses in the walls than a circular duct and has a greater tendency to deform because of high pressure within. To prevent such deformation, vertical tension posts were placed inside the exhaust duct. These posts, together with the shape transition, created a much more complicated flow than in a typical jet engine exhaust duct. This flow and the resulting heat loads on the inner surface of the tailpipe were not possible to predict, so the engineers expected that a lot of experimentation would be necessary.[34] Test cell work began approximately one year prior to first flight and quickly revealed major difficulties.[215, 232] In December 1980 a ground test of a tailpipe transition section with an operating engine resulted in a fire, which destroyed the $700,000 tailpipe.[25] The fire was attributed to

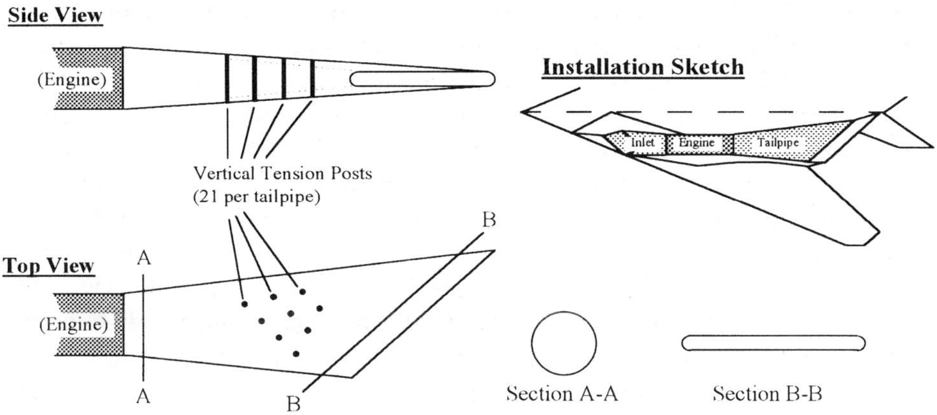

Fig. 19 F-117A tailpipe (not to scale).

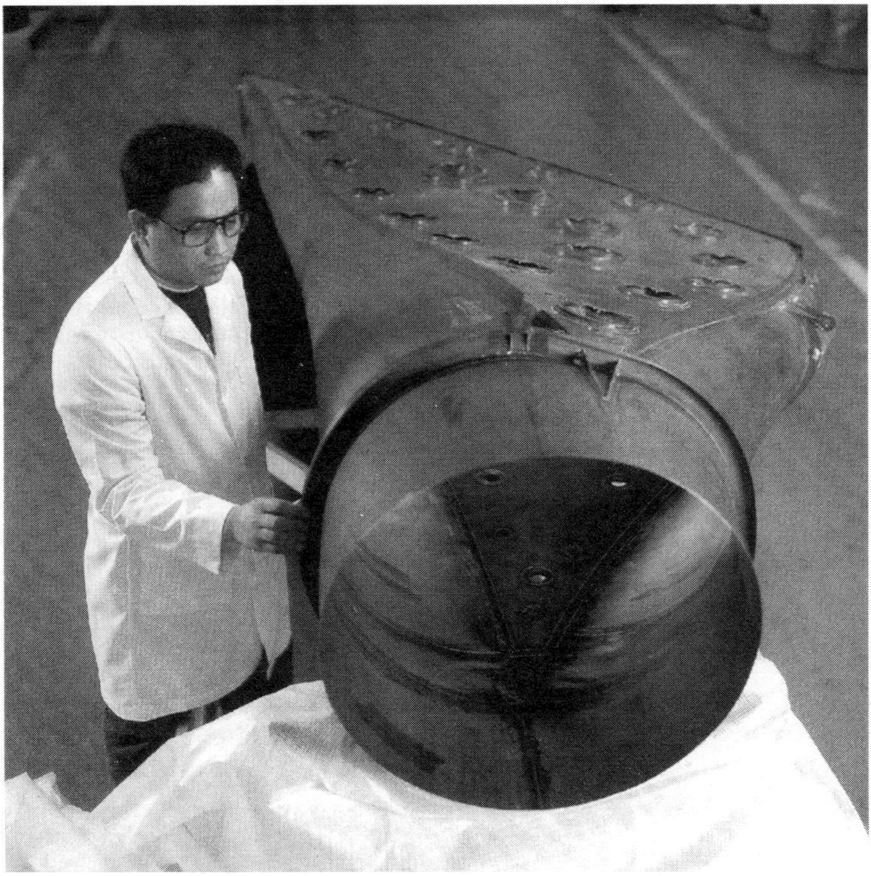

"hot spots," or locally high temperatures occurring at various locations along the tailpipe.[25] Problems with the exhaust system would continue for the duration of the full-scale development phase and beyond.

In addition to technical problems, external factors affected the Senior Trend program. Inflation rose substantially in the late 1970s and early 1980s, caused in part by economic retaliation from the Organization of Petroleum Exporting Countries (OPEC) for Western support of Israel during and after the 1973 war—the same war that had provided such dramatic evidence of the need for a stealth aircraft. The high inflation rate reduced the purchasing power of appropriated dollars. A general upswing in the aerospace industry, both commercial and military, led to increased lead times for materials, although the government at times intervened to assure that ADP parts and materials orders had adequate priority. High inflation also made it more difficult to find skilled aerospace workers.[128, 197] These problems are reported in several sources, originating not only from Lockheed but also in official Air Force correspondence. They were definitely "real" problems.

The labor problem was aggravated by the need to process security clearances for those who would be working on the F-117. Around July 1980 when the first flight originally was to have occurred the U.S. Air Force and ADP made a serious effort to resolve this problem. Very senior Lockheed officials promised priority support, while the Air Force relaxed *some* of the security constraints, for example by only requiring the highest level clearances for people who would have to know signature levels. Uncleared personnel were allowed to work on certain subassemblies in areas of the Lockheed Burbank plant outside of the Skunk Works. This enabled Lockheed to put an additional 100 to 130 people to work on the project, but it also created the need for a cumbersome, separate part numbering system that, for security reasons, had to be completely unrelated to the actual numbering system.[170, 197]

These difficulties culminated in an 11-month slip from the originally planned first flight date of July 1980. According to Ben Rich, "Missing that July 1980 deadline for the first flight wasn't the end of the world, but it made me apprehensive because I could not honestly report that the worst delays and problems were behind us." [197] In fact, at that time, final assembly of the first F-117A had not even started; it was scheduled to begin in September with the first flight still optimistically planned for December 1980.[244] In reality, the first F-117A did not reach completion until March 1981 and eventually flew on June 18, 1981.

FLIGHT TEST

TEST PLANNING

Both development test and evaluation and initial operational test and evaluation were to be performed by a Joint Test Force. This approach was similar to the one used in the Air Force Lightweight Fighter Program in 1974. As stated in an Air Force briefing supporting the FY81 budget request:

> The Joint Test Team (JTT) concept will be employed in the F-117A flight test program. The JTT will consist of contractor and Air Force Systems Command pilot/engineers plus operational [Tactical Air Command] pilots/analysts. This joint effort will eliminate unnecessary duplication and provide early operational evaluation.[90]

Air Force Systems Command participants included both SPO personnel and the Air Force Flight Test Center detachment at the test site. Because of the abbreviated program management structure, the Air Force Test and Evaluation Center* (AFTEC) was not involved.[255] In other programs, AFTEC normally acted as the independent operational test and evaluation organization; in the F-117 program, TAC was responsible for operational test and evaluation. The briefing went on to note,

> As in any new RDT&E [research, development, test and evaluation] program, lessons are being learned, this is even more significant in the development of the F-117A. Incorporation of the low observable technologies into a *weapons system* has provided unique challenges throughout the design process. Further challenges are expected throughout the test and evaluation phases of the program.[90]

These statements reflect the realistic approach and acceptance of certain risks, which were important characteristics of the F-117A's development.

---

*Subsequently redesignated the Air Force Operational Test and Evaluation Center (AFOTEC).

To achieve initial operational capability as early as possible while accommodating the inherent risks, program managers decided to accept certain limitations in the aircraft's flight envelope, weapons employment capability, and other areas.[7] Achieving an operational capability for the F-117A was therefore officially referred to as limited initial operational capability. Various limitations would be addressed through continued testing and development after limited initial operational capability was achieved.

The full-scale development test program was then planned and structured to focus exclusively on those goals and milestones that were part of the limited initial operational capability requirements:

> In order to provide focus to all engineering and flight test tasks, SPO Engineering [in conjunction with TAC and Lockheed ADP] defined the limited flight envelope to be cleared by initial operational capability and stated performance oriented goals as to what capabilities were to be achieved for each milestone.... Progress toward the achievement of these goals was briefed at each Program Management Review to Mr. Sherm Mullin and to Col Dick Scofield [the respective Lockheed ADP and Air Force program managers at the time most flight testing was being conducted. Col James S. Allen, the TAC operational group commander, was also involved in these briefings], thus keeping management aware of progress toward goals, maintaining flexibility to assign priorities to rapidly solve problems, and to enlist vendor support early if necessary.[101]

According to Mullin, "Normally all decisions were made on the spot with all three [Mullin, Scofield, and Allen] in agreement." [181] Thus the program was set up to allow resources to be brought to bear rapidly to solve problems as they arose, with Air Force (end user as well as program office) and contractor management all involved and with a minimum of bureaucratic or administrative delay. This aspect of the F-117A development is an excellent example of *managing* or *accommodating* risk, as opposed to *avoiding* risk, and illustrates a valid way to pursue aggressive, revolutionary advances in weapons system capabilities when a technological opportunity exists.

Schedules were typically optimistic and based on the assumption that everything would work. However, key officials up the chain of authority had been advised from the beginning that there were significant challenges to be overcome, and so the resulting delays did not come as surprises or threaten the program support. According to Jack Twigg, the philosophy was to "plan for, but not demand, success." [234] As the program encountered delays, the projected limited initial operational capability slipped continuously from its originally planned date of March 1982.

Both the security demands of the program and the limited facilities available at the test site required specialized logistics. Preparations for Senior Trend logistics support began in late 1979 at about the same time that construction of the first full-scale development aircraft began, which was also the same time as the first long-lead production activities were authorized. Support requirements were outlined in a staff summary sheet signed by Lt Gen Kelly Burke, U.S. Air Force deputy chief of staff for research and development on Nov. 20, 1979 [20]:

> Advanced Technology Aircraft (Senior Trend) has advanced to the point where logistics activities in support of the full-scale development and production programs are required. The initial operational capability for Senior Trend is scheduled for March 1982. Initial logistics actions by the Air Force Logistics Command must be accomplished in FY 1980 to ensure appropriate weapon system support at the initial operational capability date.
>
> Air Force Logistics Command has designated Detachment 6, 2762nd Logistics Squadron (Special), Norton Air Force Base, CA, as the logistics activity.... (This detachment currently provides logistic support for the SR-71 aircraft.)
>
> The stringent security measures currently in force within government and industry for the development and production activities will be imposed on the logistics support.... The proposed Delegation of Special Authority (DSA) and Determination and Findings (D&F) will terminate 30 September 1981 unless previously extended to provide a mandatory review point of the requirement for continued special authority.

In addition to the overall high level of security, the special provisions included dedicated airlift for repair and support. The principal airframe repair facility was at Lockheed ADP in Burbank, California; the government logistics facility was at Norton Air Force Base, California; and engine repair was conducted at General Electric in Ontario, California. Parts, engines, and airframes, therefore, had to be transported covertly between these facilities and the test site.[47] The 4450th Tactical Group was heavily involved not only in the flight testing but also in the development of training and maintenance procedures and establishing the infrastructure needed for an operational unit. Permanent living quarters for 4450th Tactical Group personnel and their families were at Nellis Air Force Base, Nevada, (near Las Vegas). This required another dedicated airlift operation using leased Boeing 727s and 737s and other aircraft with specially cleared aircrews to ferry group personnel between Nellis and the 4450th Tactical Group's operational base at Tonopah, Nevada (185 miles away) on a regularly scheduled basis.

First Flight Preparations

The five full-scale development aircraft carried tail numbers 780 through 784.* As of March 1981, Aircraft 780 (which was not equipped with full signature reduction treatments) was approaching readiness for its first flight. The First Flight Readiness Review at Lockheed ADP was tentatively scheduled for March 17–18, 1981, then rescheduled, a week at a time, for nearly every week thereafter until it finally occurred on April 27–28.[159–161, 163–169] Col Don Merkl, the program manager, was either visiting ADP weekly or living there during this period, and he sent frequent status reports to the SPO, to RDPJ, and to the test site.

Fuel leaks seem to have been the most persistent problem at this stage. They proved difficult to fix, and the leaks also delayed other necessary systems checkouts because of safety considerations with the vapor or leaked fuel.[160] Additionally, a wing tank check valve had been installed backward, and the wing tank transfer pumps would not prime consistently.[163] Merkl noted that wing fuel was not required for first flight, but that if these problems were not fixed before the first flight they would have to be added to the "fix it later" list, which was already experiencing a "snowball effect" and threatening to throw subsequent planned tests out of sequence. Merkl therefore recommended accepting the delay up-front and fixing the problems, noting that this would

*Standard Air Force tail numbers contain the year that an aircraft was put on contract. The F-117As used their own numbering system so that a sighting of an F-117A would not reveal the year it had been ordered, from which the timing and the pace of the program might be inferred. Although the initial F-117A number, 780, stood for the target first flight date of July 1980, this was not the standard serial numbering format, so an uninitiated observer would not know the meaning of the number. Subsequent F-117A numbers—781, 782, and so on—did not bear any relationship to an actual date.[223]

delay first flight "to at least May 6." His message for that week concluded, "If you don't know when to come for First Flight Readiness Review as a result of reading this message, then you understand it clearly!" [163]

Other problems included control surface tip interference caused by wing bending. The control surfaces were trimmed to eliminate this. "Oil canning" of the thin titanium firewall was experienced during a ground shake test on April 4, which required the engine to be removed in order to add a stiffener.[165] The shake test had also revealed a control/structural coupling at around 7 Hz. A simple breadboard filter was able to correct the problem, but flight simulations at Lockheed's Rye Canyon research facility showed that the new filter made the aircraft unstable at some flight conditions.[166] A double-notch filter could correct the problem without undesirable side effects, but the Lockheed controls group preferred a different approach—moving the pitch rate gyro forward, presumably to a location on the airframe where it would not pick up angular rates associated with the 7 Hz structural mode. "But this can't be tested 'til engines are in, and engines can't be in 'til fuel is fixed...." At this point there were still four fuel leaks, "all structural, three new and one being worked for the third time." [167]

The First Flight Readiness Review was finally held on April 27–28, 1981, at Lockheed ADP. The formal review included tailpipe inspection procedures. These would consist of a detailed inspection prior to the high-speed taxi test, and every 10–25 flight hours thereafter. This was considered to provide sufficient flexibility in the test program without compromising flight safety (although in practice, visual inspections were also conducted before every flight). A few action items remained. ADP had previously cut out a portion of the seat pan (a plastic part at the base of the ejection seat) to eliminate an interference problem with the control column. Merkl required ADP to obtain a written statement from the ejection seat vendor that this did not have any safety implications. Additionally, some ambiguities needed to be resolved regarding the results of a recent braking test, which had resulted in a fire. The ambiguities concerned whether the fire had occurred during or after the test and if the fire resulted from differences in configuration between the tested article and the "real" brakes.[169]

To ensure that the aircraft would not be delivered in an unsafe or incomplete condition, the following acceptance procedures were required[169]:

1) A responsible Lockheed ADP official would sign a statement that the aircraft was safe and ready for flight.

2) A DD250 [aircraft acceptance form] would be prepared with an attachment indicating all "missing" equipment and planned dates for installation of same.

3) A SPO Question and Answer section would be used to verify that all "required before flight" items were closed.

4) Col Merkl would sign the DD250 on behalf of the government.

5) An immediate contract modification would then provide the aircraft back to ADP as "government furnished property."

6) The aircraft would then be placed under the responsibility of the Joint Test Force director.

The review was considered completed at around noon on April 28, and Aircraft 780 was immediately transported to the test site. An engine run was performed that evening, resulting in an event that Merkl described as a "fire" and Alan Brown of ADP described as an "exhaust system overheat problem." Following a 3 min engine run, flames were observed in two areas. The trailing edge just outboard of the nozzle caught fire because a seal between one of the protective "bricks" and the nozzle side member had failed; an internal fire had occurred because of fuel accumulation in the pan area formed by the structural skin just forward of the nozzle. Both fires were extinguished within 14 s without serious damage to the aircraft. Required fixes were identified during an ad hoc "post-fire session" of the First Flight Readiness Review.[32, 169]

The various issues were worked, and on June 12 "final" preparations were made for the high-speed (165 kn) taxi test, scheduled for June 13, which would be the last ground test prior to first flight. However, cracks were discovered in the right-hand inlet duct, forcing the test to be postponed to June 14. An engine run on June 13, involving 10 min at maximum thrust to obtain platypus temperature data, caused some damage to the nozzle area. This was repaired, and another engine run on June 14 was successful. However, the taxi test was postponed because of high winds. The test finally occurred on June 15. During the next two days, the test team made final flight prepara-

tions, which included instrumentation work and repair of one remaining fuel leak. Meanwhile, Aircraft 781, still undergoing assembly at Burbank, was fueled for the first time on June 16, and seven leaks were discovered.[1, 2, 3]

The foregoing discussion provides some insight into the kind of difficulties that were encountered in preparing the first example of an operational very low observables aircraft design for flight. The unusual shapes and treatments of the aircraft and the exhaust system resulted in many thermal, structural, and material problems. These problems and many new ones would persist throughout the flight test phase of the program. In retrospect it was a wise choice to pursue the development of the lower risk ATA "A" rather than the larger ATA "B," with its much higher takeoff weight and hotter, partially afterburning engines.

Attendance at the first flight was strictly controlled. In early March Col Ken Staten (at RDPJ) had reported to Col Don Merkl (at the SPO), that

> Secretary Orr [Secretary of the Air Force] and General Burke [DCS/R&D] have agreed that they will not attend and will not favorably view other requests. I am complicating the problem for any potential attendee by refusing to supply definite dates [something he could do with complete honesty, in view of the technical difficulties that were experienced]. Please forward a list of those senior contractor, AFFTC, and ASD people that will/may attend. I stand ready to provide any assistance you need in providing higher authority for disapprovals.[218]

Maj Gen Lawrence Welch (Tactical Air Command director of operations) had also agreed not to attend. This reflects one of the key ingredients to successfully running a "streamlined" program: a strong commitment on the part of the government to keep the program small and "outsiders" out. Several high ranking "insiders" in this case set the correct example by not attending the first flight.

First flight reporting procedures were established ahead of time and followed the general philosophy of short and direct lines of communication all the way up the chain of command. Col Merkl would notify by secure communication the SPO at ASD and AF/RDPJ at the Pentagon. The SPO would notify the ASD commander (ASD/CC), and AF/RDPJ would notify those in the Washington, D.C., area, specifically the under secretary of defense for research and engineering, the deputy secretary of defense, a very small cadre of cleared individuals within the Office of Management and Budget, and the relevant Congressional committees providing oversight of the program: the House and Senate Armed Services Committees, and the House and Senate Appropriations Committees. A small number of people at Tactical Air Command headquarters would also be notified, specifically the director of operations, the vice commander, and the commander. All of these notifications would be made by secure communication or cleared courier. The notification would include a brief written summary with the pilot's name, time, and duration of flight, test events completed, difficulties, engineering assessments, and near-term plans.[161, 168] In addition, RDPJ requested a short film clip and a few photos. "Such will be necessary to satisfy the 'hungry mob' and will be beneficial to the program."[218] These would have to be suitable for appropriately cleared viewers.

Some concern over the accuracy of the faceted air data probes existed in the area of flight safety. As noted earlier, the F-117A was designed to be statically unstable in the longitudinal (pitch) axis, and therefore required active feedback of aircraft attitude and rate information to the flight control computer. Feedback in the longitudinal axis consisted of pitch angle, pitch rate, and normal acceleration. Although not critical for safety of flight, the control system also used feedback in the lateral/directional axes to provide stability augmentation. Angle of attack α and angle of sideslip β were determined from the air data probes. However, unlike conventional probes that have rounded tips to ensure smooth airflow, the probes on the F-117A are faceted, with a pressure port on each facet, as shown in Fig. 18.

Typical of the first examples of any new type, Aircraft 780 and 781 carried an additional, conventional flight test pitot-static noseboom mounted on the centerline of the aircraft to ensure that accurate airspeed, attitude, and altitude information were obtained. However, this flight test noseboom provided calibration data only, and was not wired into the flight control system.[175, 212]

Hence, the automatic flight control functions were entirely dependent on the data provided by the new, faceted probes. According to Skip Hickey, who had been the Air Force flying qualities engineer for Have Blue and was chief engineer for Senior Trend, "There was some concern of the accuracy of these readings; therefore, it was decided to make the first flight with a statically stable air vehicle" so that the pilot could safely fly without the alpha feedback.[101]

Therefore, for the initial flights, adding ballast in the nose of the aircraft created positive static stability. The alpha and beta feedback, as well as the aileron–rudder interconnect, were then routed through individual switches so that the pilot could select them as desired. The other feedback data —rates and accelerations, all provided by conventional, proven inertial sensors—were left connected. The intent was to take off and climb with alpha, beta, and aileron–rudder interconnect disconnected, and then activate these channels one at a time at an altitude of 15,000 ft to evaluate the control responses. "With some marvelous foresight, rubber tubes were put on the top of the switches to these channels to extend them so that the pilot could rapidly engage them with a swipe of his hand if required." [101] Also with excellent foresight, simulations had been conducted on the ground at Rye Canyon and in the air using the Air Force Flight Dynamics Laboratory's NT-33 variable stability test aircraft* of various failure and degraded stability conditions. The pilot was, therefore, prepared in advance as well as possible for a variety of problems that might occur.[54–56, 101, 175]

First Flights

The first flight took place on June 18, 1981, with Lockheed pilot Hal Farley at the controls. Nose wandering was observed immediately after takeoff: first 2 deg right, 3 deg left, then 4+ deg right, at which point the pilot quickly turned on the yaw feedback, which alleviated the immediate problem. The simulations of degraded directional stability that helped the pilot recognize the problem and the makeshift switch extenders that allowed the pilot to activate the feedback loops without taking his attention away from flying the aircraft paid off. Pitch feedback and aileron–rudder interconnect were activated later in the flight, as planned, and worked properly. Pitch/airspeed control were excellent, although the airspeed indicator read low by up to 8 kn at some airspeeds. Annoying environmental control system surges were experi-

---

*The Calspan NT-33A was a specially modified Lockheed T-33 trainer used for simulation and flying qualities research from 1958 until the late 1980s. The NT-33's stability characteristics, control system, and displays could be altered to artificially simulate other aircraft. It was used to evaluate and to familiarize pilots with the flying qualities of several new aircraft prior to their first flights, including the F-15, YF-16, YF-17, F/A-18, and F-117A. The NT-33A has recently been replaced by the Lockheed/Calspan NF-16D VISTA variable-stability aircraft.[104]

enced along with thumping of the blow-in auxiliary inlet doors. Exhaust nozzle temperatures were higher than desired. There was an intermittent canopy unsafe light, possibly caused by the environmental control system surging, which led to early termination of the flight. The airplane handled well during approach and landing. Total flight duration was 14 min.[54]

All of this information was promptly reported up the oversight chain, as discussed, and program managers noted that the canopy warning and nozzle temperature problems would have to be corrected before another flight would be made. (Because of its historical value, the three-page test pilot summary report is presented in Appendix F.)

Col Ted Bowlds (current commander of the U.S. Air Force Rome Laboratories), as a captain at that time was a member of the Senior Trend Joint Test Force. Capt Bowlds was responsible for preparing the first flight film clip to be sent to Washington:

> Getting to that short clip (about 1.5 min in length) took some effort. The test site did not have the capability to develop motion picture film. Rather, we worked an arrangement with Nellis Air Force Base to use their equipment. Ultimately, we trained our photo shop personnel on how to develop film, collected the footage from the first flight, flew to Nellis, developed the film, with various clipping built the movie, and met Col Merkl at the airport so he could deliver it to Washington.[31]

During the next several flights, basic performance, flying qualities, and functional checkouts continued, and minor problems were worked. Concurrently, there were continuing efforts to improve the aircraft's lateral/directional handling qualities. Although these efforts were not entirely successful, the initial airworthiness tests were considered to be complete after the sixth flight. Two other pilots then had an opportunity to fly Aircraft 780. These were Col Skip Anderson, the first Air Force pilot to fly an F-117A, and David Ferguson of Lockheed, who later became Lockheed's chief test pilot for the YF-22 Advanced Tactical Fighter concept demonstration test program. The first 10 test flights are summarized in Table 6.[9, 10, 54–61]

Following its 10th flight, Aircraft 780 was grounded for scheduled maintenance and inspections. Larger tail fins were also added to improve the aircraft's directional stability. Aircraft 780 first flew with the larger tails on Oct. 21, 1981. These were interim fins, with added area built onto the same structure, as a quick fix prior to the analysis and testing that would be required to design the final production fins. The interim enlarged fins were also retrofitted to Aircraft 781 and incorporated on all subsequent full-scale development aircraft.[93, 207]

Security procedures during early F-117A flight testing were similar to those used during Have Blue. Flights were made only during the first hour or so of

Table 6 Summary of first 10 F-117A test flights, Aircraft 780

Flight No.	Date	Flight	Pilot	Purpose/comments
1	June 18	14 min	Farley	*First flight.* All major objectives accomplished. Directional stability less than expected. Pitch and roll okay. Flight shortened due to canopy unsafe light. Nozzle temperatures also too high.
2	July 1	39 min	Farley	*Flying qualities and functional checks.* Landing gear operation and dynamics excellent. Directional stability still too low. Good acceleration; T-38 chase plane required half afterburner to keep up. Nozzle temperatures better.
3	July 9	10 min	Farley	*Handling qualities, loads and flutter.* Terminated early due to lateral/directional oscillation.
4	July 15	51 min	Farley	*Stability and control, functional checks.* Investigated cause of directional oscillations. Emergency gear extension unsuccessful (did not indicate locked).
5	July 23	1h, 5 min	Farley	*Envelope expansion.* Very successful. Expanded envelope to 330 KEAS, 3.0 *g*, 10 deg α). Performed rolls, windup turns, formation. Emergency gear extension still unsuccessful.
6	July 25	1 h, 27 min	Farley	*Flutter, flying qualities, loads, functional checks.* More yaw damping needed. Flying qualities otherwise excellent. Several engine crossbleed airstarts unsuccessful. Emergency gear extension still unsuccessful. Initial airworthiness tests complete.
7	July 28	1h, 2 min	Anderson	*Pilot familiarization.* Yaw oscillations were disturbing. Otherwise good flying qualities. Simulator fidelity is poor.
8	July 30	1 h, 18 min	Anderson	*Performance and functional checks.* Poor handling in turbulence.
9	Aug. 1	1h	Ferguson	*Pilot familiarization.* Flying qualities very pleasant.
10	Aug. 6	49 min	Ferguson	*Flying qualities.* Shortened due to Soviet satellite cover time and a late takeoff due to conflicts with another program.

daylight, and strict measures were used to prevent any unauthorized sightings of the aircraft.[34]

FULL-SCALE DEVELOPMENT TEST PROGRAM

Planned and actual testing during the full-scale development phase of the Senior Trend program are illustrated in Figs. 20 and 21, both of which come from a briefing given shortly after initial operational capability by Lt Col Roger Moseley, then-commander of the Combined Test Force (CTF).[179]

Aircraft 780 was used for most of the testing related to the basic air vehicle—loads, flutter, takeoff/landing, and high angle of attack. In keeping with its purpose, this aircraft retained its flight test noseboom for the duration of the test program.[212] Aircraft 780 was grounded from April through June 1982 because of a battery acid spill. "The first time we flew [an F-117A] inverted, the battery acid spilled out," causing damage to the aircraft, recalled Paul Kaminski.[116] It was grounded again during October and November 1982 for scheduled 100 h maintenance and inspections, after which it resumed testing up until initial operational capability.[179, 211]

The second F-117A, Aircraft 781, passed its First Flight Readiness Review at Lockheed ADP on September 2, 1981. On September 24, 1981, Aircraft 781 made its first flight, but after only four sorties it too was brought down for work, which included the retrofit of the larger, interim design vertical tails already noted.[93] A "production" nose was also added at this time, which included an asymmetric four-probe air data system versus the five-probe early flight test configuration, and provisions for a fully functional infrared acquisition and designation system (IRADS), to include a FLIR unit and a DLIR unit, although these were not yet installed. Aircraft 781 first flew with the production nose and larger interim fins on December 1, 1981, but this flight was aborted due to DLIR screen failure shortly after takeoff, caused by acoustics in the empty IRADS bay. This was repaired, and several flights were made.

On December 18, Aircraft 781 was again grounded for approximately one month for RCS coating work. Then, after one checkout flight, Aircraft 781 made the first RCS flight in the Senior Trend program on January 23, 1982, flown by Air Force pilot Lt Col Roger Moseley. This was Moseley's second flight in an F-117A.[211] Following a short series of RCS test sorties through early February, 781 was grounded until June for trapeze (weapons carriage: the F-117A has two retractable trapezes to carry the two weapons inside the bomb bay) and inlet and pitot system de-icing work. The aircraft was then used primarily for weapons separation and anti-icing testing and development, as well as some of the flying qualities and performance testing, for the remainder of the full-scale development test program.[179, 212]

The first flight of Aircraft 782, on December 18, 1981, was aborted because of high platypus temperatures. Following resolution of this problem and the successful accomplishment of airworthiness and functional checkout, 782 was then used for acoustics (probably IRADS bay acoustics because of the timing with respect to the recent screen failure on Aircraft 781, but weapons bay acoustics was also an important issue) and navigation system testing. This aircraft made the first F-117A night flight on March 21, 1982.[211] It was then grounded for trapeze work and subsequently used for weapons delivery testing, further acoustics work, initial operational testing and evaluation, and extensive full-up avionics testing of the IRADS system.[179]

Aircraft 784 was the next to fly, on April 10, 1982. This aircraft was also used for navigation/autopilot and avionics development and evaluation, and in particular was used for acoustics work, which continued right up until initial operational capability.[179, 211]

Aircraft 783 had been delivered in January or February 1982 but remained on the ground until July while undergoing RCS preparation. It made its first flight on July 7, 1982, and following its initial airworthiness testing was briefly grounded for fuel work. This may refer to leaks, which were particularly critical once the radar absorbent material (RAM) coatings had been applied, because the fuel had a tendency to attack these coatings. As with many troubleshooting and maintenance tasks, the work was labor-intensive because some of the RAM coating would have to be removed to locate and repair the leaks, and reapplied following the completion of the repair. Aircraft 783 was then used as the primary RCS and infrared signature test aircraft through early 1983, when it was grounded for systems work. Then, from summer of 1983 through initial operational capability demonstration it was used for avionics integration and further signature testing.[179, 211]

The first production F-117A, Aircraft 785, was also originally intended to participate in the test program. However, it was destroyed on its first takeoff attempt at the test site on April 20, 1982. Pitch and yaw rate gyro input to the flight control computer had been cross-connected, causing complete loss of control immediately upon rotation for takeoff. Aircraft 785 had been intended for use in maintenance training. The next, Aircraft 787, first flew in August or September 1982 and was used by the test force for initial operational test and evaluation. Due to various delays, Aircraft 786 was not actually delivered until after Aircraft 787. Aircraft 786 and all further production aircraft were delivered to the operational squadrons.

Throughout the F-117A's test and evaluation program, the Combined Test Force (CTF) bore a heavy responsibility, often beyond that normally associated with a more conventional test and evaluation effort. According to then-Capt Ted Bowlds,

> From my viewpoint, the CTF was responsible for not only the day-to-day testing of the F-117A, but because of the high degree of concurrency, the CTF seemed to take on a higher degree of engineering responsibility. Most of the daily dialogue with the contractor over design aspects were handled by the CTF with oversight from the program office. Again, the main reason was the limited number of people involved in the program (and the CTF's proximity to the contractor).[31]

As an example,

> Human factors testing was accomplished in a somewhat unorthodox fashion. A production representative flight test aircraft was positioned outside at night with all outside lighting turned off. Test pilots were se-

> quenced through the cockpit and asked to comment on various aspects of the cockpit lighting. The results of this exercise were then used to "tune" the interior cockpit lighting. I felt that this was unique because when the program office eventually obtained a human factors engineer, the CTF was told the original test was invalid for a variety of reasons and Lockheed was placed on contract to determine the correct cockpit lighting. Bottom line, the results were the same.[31]

Lt Col Stan Siefke provided an interesting perspective on the role of personnel assigned to the CTF during the early days prior to F-117A's Initial Operational Capability,

> When I first arrived as a captain, I was assigned to the team of Air Force and Lockheed engineers responsible for all performance, flying qualities, aircraft loads, flutter, and stores separation testing. Lockheed team members at that time were generally assigned to a specific tail number and were responsible for the actual aircraft preparation for flight test overseeing duties ranging from fuel onloads and preflight checks to post-flight instrumentation calibrations. Their experience levels ranged from seasoned flight test engineers from previous Lockheed programs as well as Rockwell hires experienced on the B-1 program to brand new young aeronautical engineers fresh out of college, but all were a dedicated lot realizing the importance of their positions in the Combined Test Force (CTF) and the importance of the development of this aircraft. Early team members and their assigned aircraft were Hap Roschon, A/C 780; Dennis Fernandez, A/C 781; and Dan Fuller, A/C 782.
>
> Air Force engineers were assigned to oversee general areas of expertise and consequently saw flight test action and mission preparation on many of the CTF aircraft. I was initially assigned the validation of the aircraft's basic performance capabilities (takeoff, climb, cruise, turn, descent, and landing performance testing) as well as helping on the aircraft loads and flutter testing and stores separation testing. All U.S. Air Force flight test engineers were experienced graduates from the U.S. Air Force Test Pilot School and almost all continued on to become key decision makers and managers in the classified low observables world.... By the time Initial Operational Capability (IOC) approached in October of 1983, I was responsible for all flight testing of Flight Control Computer (FLCC) modifications (high angle of attack) testing for the '-7 FLCC' and beyond, finishing up all the performance testing of the aircraft, designing, and conducting all-weather icing testing on A/C 781, and completing the arresting hook testing on A/C 780. Lt Col Moseley, who was the CTF commander at the time (having replaced Lt Col Paul Tackaberry) put me in charge of three critical aircraft-related 'watch items' required to be completed prior to the declaration of initial operational capability. As a matter of fact, in one day I conducted an early morning arresting

> hook engagement test with Lockheed test pilot Dave Ferguson in A/C 780, a mid-morning high angle of attack test with U.S. Air Force test pilot Jon Beesley in A/C 781, and an early afternoon all-weather icing test on A/C 783 with U.S. Air Force test pilot Lt Col Roger Moseley.[212]

Fig. 20 shows the extensive post-limited initial operational capability testing provided even on the "original" flight test plan. All testing related to non-level weapons delivery modes, nuclear certification, and air-to-air missile employment, which were "desired" but not "required" capabilities, was deferred until after the planned limited initial operational capability date.[7, 101, 179] Initial operational test and evaluation was also planned to continue after limited initial operational capability (this was fairly unusual; initial operational test and evaluation in a typical program is considered to end at the initial operational capability) along with continuing avionics development. Test activities subsequent to limited initial operational capability are discussed in Chapter 6.

According to the initial plan, 782, 783, and 784 were to have been used for RCS testing during most of the period prior to limited initial operational capability. In reality, as seen in Fig. 21, only Aircraft 783 was used for this

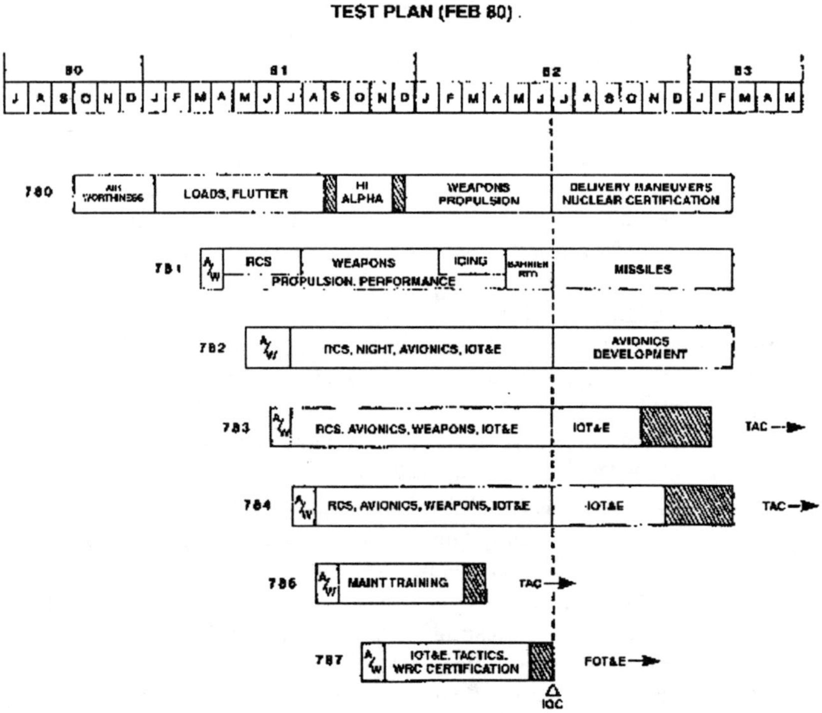

Fig. 20 Full-scale development flight test plan as of February 1980.[179]

purpose. Aircraft 781, as planned, was used for RCS testing but only for a brief period very early in the test program, probably to provide an initial "quick-look" verification that the F-117A would be able to meet its signature goals. There are two likely reasons that only one aircraft was used for RCS testing. First, the effort of maintaining three aircraft in a low-signature condition was probably determined to be prohibitive. Second, enough other difficulties with avionics and weapons integration arose to demand the full-time use of Aircraft 782 and 784 for those purposes.

TECHNICAL CHALLENGES

Of particular significance, when comparing Figs. 20 and 21, is the large amount of aircraft downtime actually experienced, compared to the very small amount planned, illustrating the optimism of the original schedule. The various problems and their impact as reported from the point of view of the Combined Test Force, are summarized in Table 7.

The most critical and persistent of these deficiencies were with the IRADS, air data probes, inlet grids, the fins, tailpipe, and landing gear/tail hook.

IRADS. The infrared acquisition and designation system (IRADS) problems concerned both the system itself, and the screens over the apertures. The

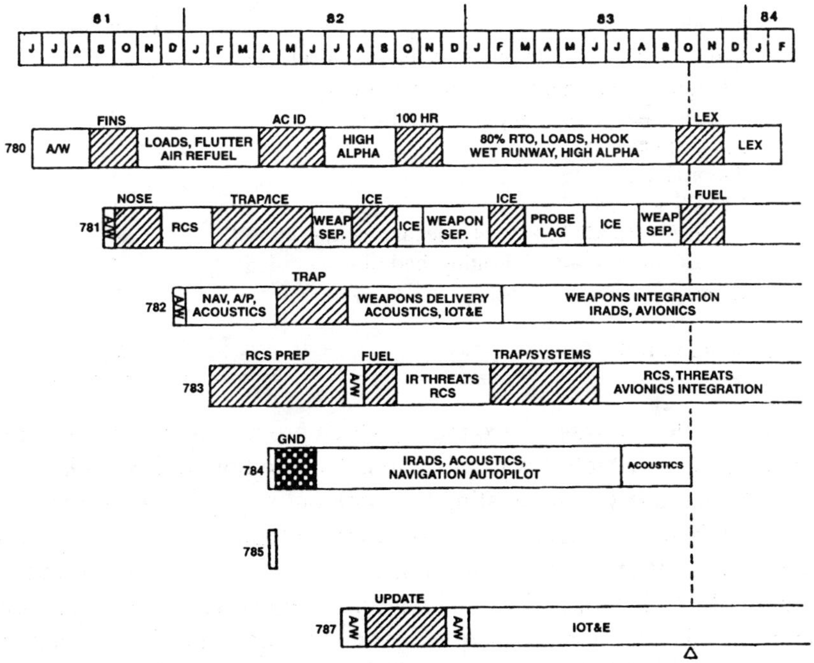

Fig. 21 Actual full-scale development flight test program.[179]

Table 7 Deficiencies encountered in the F-117A full-scale development test program[179]

Deficiency	Impact (Aircraft test months)
Basic aircraft design deficiencies	(total 22)
Initial slip in aircraft first flight (optimistic schedule and "minor" design problems)	—
Fuel leaks	4
Canopy jettison	3
Landing gear (downlock pins and drag links)	5
IRADS boresight	6
Hook (six calendar months impact)	1
Aircraft producibility	3
Wing root bending strength	—
Quality control	(total 19)
Battery acid spill	3
Airframe electrical connectors	3
First production aircraft—gyro wiring (caused crash)	5
Hydraulic and fuel plumbing preload	2
Ejection seats	1
Production grids/coating	3
Flight control electronics	2
Technology hurdles	(total 25)
Directional stability (really more of a basic airplane design mistake than a technology hurdle)	3
Reduced elevon control power at aft c.g.	2
Exhaust gas leaks	5
Tailpipe life	1
IRADS screens	1
Acoustics (IRADS bay) (14 calendar months impact)	4
Air data probes—RCS, heating, and flow	4
Coatings	5
Total up to IOC	**65**

initial screen design was etched metal, which proved unable to withstand the acoustic loads to which it was subjected. As noted by Ted Bowlds, "the design of the screens was interesting to watch but was quite frustrating. About once a month, the engineer responsible for developing the screens would display his latest attempt at etching. For the longest time, little progress was being made (we were always just one week away from solving the problem)."[31] In addition to adequate structural strength, the screens had to provide a sufficiently benign acoustic environment in the bays for the FLIR and DLIR to function properly, allow infrared and laser radiation to pass through while blocking radar frequency radiation, and avoid ice formation. As noted earlier, the DLIR screen failed the first time a "production" nose was flown in December 1981.

Windows were also under consideration. According to Sherman Mullin, who became Lockheed's F-117A program manager shortly after that failure, "To solve the IRADS aperture problem a new subcontractor had been selected to develop a solid state crystalline aperture made of zinc selenide." In March 1982 Mullin and his program team reviewed this issue in depth. Not one usable window had been built. The estimated production cost [of the windows] was $200,000 each ($400,000 per aircraft). Mullin decided to cancel the contract. The following weekend, at home, Alan Brown designed the wire grid screen, using piano wire. He selected the wire diameter, the wire spacing and created the basic mechanical design. It allowed the infrared and laser energy transmission with very little attenuation and acted like a flat facet at most radar frequencies. A prototype was built and rapidly tested. It worked, so it was put into production as an "interim solution," in conjunction with acoustic treatments to the interior of the IRADS bay.[175,181] The "interim" wire screen design is still in use for the IRADS apertures.

Problems with the IRADS system itself included aligning and merging the image from the FLIR turret with the image from the DLIR turret on a single display. The FLIR and DLIR turrets had identical hardware, except the DLIR was mounted upside down, so the DLIR image had to be inverted electronically. This caused some problems. There were also boresight problems and difficulties handing off the target from the FLIR to the DLIR. A special team was formed at Lockheed to work these problems, and approximately one year of intensive effort was required to overcome the various IRADS issues.[215,236]

The pilot workload required to lock on to targets was very high, requiring the use of the autopilot when attempting a lock-on.[211] This problem was not fully resolved during full-scale development, but only through subsequent upgrades, which are discussed in Chapter 6.

Air Data Probes. Because the F-117A is statically unstable, as noted previously, accurate data from the pitot probes was flight-critical. According to Lockheed, "Requirements forced upon the aircraft by the low-observables engineering group made the design of the probes and their deicing system a serious challenge. Many different designs were tunnel tested before an acceptable one was found," due to the seemingly divergent requirements of signature control and airflow quality.[175] The first wind tunnel test of a full-scale probe occurred in February 1979. A "production version" was tunnel tested in mid-1980.[232] However, further changes would be necessary.

The need for probe heating to prevent ice formation had been realized from the start, but it took a long time to develop and integrate a suitable heating system.[34] For this reason, the first flights were made with unheated probes.[232] When heated probes were tested in flight, the heating caused problems with probe deformation and/or cracking of the RAM coatings on the sides of the probes.[31] The RAM coating on the tips of the probes experienced a different problem: rain erosion. Eventually the tip coating was abandoned, and the bare metal surface (initially a copper alloy) was used.[34] Even the metal tips had to be inspected periodically, and the probes replaced if there was damage from rain erosion or other sources.[248]

Finally, packaging the nine pressure tubes per probe, the electrical wires, and the heating elements within the physical structure of the probe was a challenge. The original pressure tubes proved to be too thin, according to Stan Stiefke:

> The small diameter tubing that was used to convey the pressure information to the actual sensors introduced a phase lag in the pitot-static system that eventually coupled into the Flight Control Computer (FLCC) leading to a slow directional "wobble" when the aircraft was flown at high altitudes. This only added to the now famous "wobbly goblin" descriptions of the F-117A; however, once the pitot tube pneumatic lines were increased in size, the pressure lags disappeared and the FLCC got the correct information at the right time and was able to maintain the "pointy end going forward" at all times without oscillations.... Following upgrades to the flight control computer, larger tailfins, and solutions to the probe lag problem, the F-117A has, within its own flight envelope, "flying qualities just like the F-15," according to pilot comments.[212]

As indicated in Fig. 21, testing of the larger diameter pressure tubes was accomplished with Aircraft 781 during March, April, and May 1983.

Throughout the development effort, as different designs were tried, "... the pitot tubes were each hand built by a small number of individuals at Lockheed. This took a long time...."[31] The program could not have gone forward without a resolution to the various probe issues. An acceptable probe design was not achieved until shortly before initial operational capability, and at least two additional versions have been tested since then. These were a beryllium tip, developed in 1985 to provide a harder surface than the original copper alloy, and a "close tolerance tip," tested in late 1986 and incorporated into production aircraft starting in early 1988.

Inlet Grids. Considerable anti-icing work and RCS testing were required for the inlet grids. "The inlet grids not only looked like a giant ice cube tray, but acted like one, as well."[175] Metal leading edge heating elements were unacceptable for RCS. A latex de-icing boot system, similar to that used on the wings of general aviation aircraft, was tried, but it's performance was "awful," according to Alan Brown.[34] A wiper system with de-icing fluid was ultimately adopted, along with lights that could be turned on to allow the pilot to visually check the condition of the grids while in flight.

Apart from the icing issue, airflow quality through the grids was not a significant problem. Although the grids created some pressure drop with consequent thrust penalties, they actually provided very uniform pressure across the duct resulting in minimum distortion at the compressor face. Furthermore, the inlet grid pressure recovery was nearly constant for all angles of

attack and sideslip normally encountered with the F-117A, due to a very clever design of the grid leading edges.[34, 175]

Tail Fins. The directional stability problem noted previously about the first flight proved to be very persistent. By the second flight, even with yaw feedback activated, the closed-loop directional stability was less than satisfactory. The yaw angle gain was therefore increased for the third flight. However, this led to a rapidly growing lateral/directional oscillation as soon as the landing gear was retracted. Flight control engineers on the ground quickly identified the problem: yaw gain was programmed to change with gear position, with a higher gain value being used when the gear was up. They promptly advised pilot Hal Farley to lower the gear, returning the flight control system to a lower, safer yaw gain.[55, 56] Rather than solving the problem, the higher yaw angle gain had made the system dynamically unstable. The rudders automatically overcorrected for any yaw error, thereby causing the aircraft to yaw the other way, which would then be overcorrected again, etc. Following some simulation experiments at Rye Canyon facility, a higher yaw *rate* gain (as opposed to yaw *angle* gain) was adopted, which restored the dynamic stability somewhat.[57]

However, completely satisfactory lateral/directional characteristics were not achieved through control law modifications alone during the first 10 flights, as evidenced by Farley's comments on the sixth flight and Col Skip Anderson's comments on the seventh flight (see Table 6). In the meantime, considerable effort had been devoted to assessing the underlying cause of the directional instability. One of the flights had been made with tufts on the fuselage to observe the airflow and determine if there were any anomalies. Also, according to Skip Hickey, "A thorough reevaluation of the wind tunnel data was undertaken. It was found that the sting mount used to test for lateral directional force and moments greatly distorted the aft end of the model. The corrections for this distortion were incorrect, and the air vehicle was more directionally unstable than predicted. Furthermore, the directional data was not corrected for flowing inlets.... As a result of these analyses, the vertical fins were increased," to restore the levels of basic air vehicle directional stability that had been originally assumed.[101]

Larger fins were retrofitted to Aircraft 780 while it was on the ground for scheduled maintenance and inspections following its 10th flight. Aircraft 781 also received the larger fins following its first few flights. Both of these aircraft were flying with the modified fins by late 1981. This was a "quick fix" to keep the test program going in the short term. Initial analysis indicated that the existing structure was sufficient to support the larger planform area (although the program did not have a complete fatigue test article at that time).[34] Starting in the fall of 1981, program managers considered options for a more permanent solution to the directional stability problem and laid out budgets and test plans. Winglets (vertical surfaces at the wingtips) were proposed as a possible way to augment the aircraft's directional stability, but were quickly rejected.[206, 207] As it turned out, the interim fin remained the only fin until 1985 when a fin failure occurred in flight.[34, 232]

Tailpipe/Platypus. Exhaust system problems persisted from well before the first flight until after the initial operational capability demonstration. The heat shielding around the nozzle experienced problems; the protective high temperature ceramic bricks and small seals had a tendency to come off, allowing hot exhaust gas to come in contact with unprotected parts of the airframe. The platypus deck, where the lower lip extended aft of the nozzle exit, was found to deform due to the exhaust heat. The solution, credited to Lockheed's Henry Combs, was a shingled design for the platypus deck in which panels could slide over one another as they expanded.[175] Structural integrity of the tailpipe itself was also an issue, and more exhaust gas leaked.[179] In addition, the unusual exhaust nozzle shape imposed unexpected loads on the rest of the airframe.[175] Platypus temperatures were often higher than they were supposed to be, as noted on the first flights of Aircraft 780 and 782. This was a frequent source of concern throughout the flight test program[9, 10, 54–61] and precipitated the design of an alternate tailpipe. In December 1981 the ASD program manager, Col Richard Scofield, noted that the F-117A "will need [the] alternate tailpipe if we want > 200 h reliable life; decision on alternate design [is required by] mid-January 1982."[206] Nevertheless, he was forced to temporarily suspend the alternative tailpipe effort due to funding constraints.[207] Eventually, a tailpipe liner developed by General Electric was adopted.[101]

Problems continued for the duration of the full-scale development phase. For example,

> On Friday June 24, 1983, we decided to withhold the Senior Trend fleet from flying. Cracks had been found in tailpipe support posts during Lockheed's investigation of a recent tailpipe failure.... Our action plan is: a) Remove all tail pipes, extract and X-ray all posts (21 per tail pipe) for defects.... If all goes well, the entire fleet should be cleared for flight within two weeks.... The SPO and the contractor are assessing if there are any impacts to flight test, production, and IOC [initial operational capability] (Aug. 31, 1983).[150]

The posts referred to were vertical pieces in the tailpipe itself, which prevented the top and bottom surfaces of the tailpipes from bulging outward (as shown in Fig. 19). All five full-scale development aircraft and several production aircraft were flying by this time. Removing, inspecting, and replacing 21 posts per tailpipe, two tailpipes per aircraft, must have been an interesting project, to say the least.

Even at initial operational capability, which was actually achieved on October 28, 1983, the tailpipe was still a cause for concern, and the 4450th Tactical Group commander had doubts as to whether the tailpipe durability was sufficient to safely accomplish extended-duration worldwide deployment flights.[8] Exhaust system development and improvement has been under way continuously through the present time, and is projected to last at least until 1998.[232]

Tail Hook. The tail hook system was not designed to handle the increased loads that could occur during arresting barrier engagements under high-speed emergency abort conditions at the departure end of the runway. At maximum takeoff gross weight (52,500 lb), the aircraft rotation speed was in excess of 150 KIAS with a lift off speed of over 175 KIAS. "Additionally, the damping and hold-down system used to keep the hook in contact with the runway in order to engage the arresting cable was designed using poor assumptions," according to Lt Col Stan Siefke, who as a captain was the Combined Test Force flight test engineer responsible for the arresting hook tests.[212] This not only increased the possibility of failing to engage but the bouncing tailhook also caused damage to the underside of the aircraft and particularly to the RAM coating. "The end result was that a combined Lockheed and Air Force team pitched in with the aid of real flight test data to redesign the hook system using the 'fix–test–fix' philosophy."[212]

Full-Scale Development Cost Increases

The various technical difficulties were accompanied by cost increases. In fact, as early as August 1979, shortfalls had been identified in both full-scale development and production funding through FY81.[155] Senior Trend apparently operated with considerable freedom to transfer money between research, development, test, and evaluation (the budget category that covered all full-scale development activities) and production but still required certain outside approval to change the total budget. Seeking outside approval carried some risk of drawing attention to the program; so the program management always tried to live within its total appropriated dollars, waiting for future appropriations to "fix" any net shortfalls. A frequently used solution to short-term full-scale development funding problems was therefore to move money from production to full-scale development. This had to be done repeatedly as full-scale development costs increased steadily over the course of the program.[30, 91, 140, 155, 156]

With the flight test activity picking up in late 1981, and the consequent emergence of a variety of "new" technical problems, the funding problem

became too big to solve with transfers from Senior Trend production alone. Col Scofield notified RDPJ and Air Force Systems Command headquarters

> ... that we cannot comply with the direction provided in the Nov. 13, 1981 Senior Trend PMD [program management directive]. As a result of the final FY82 Senior Trend appropriation...and a recently identified prime contractor cost growth in FY82 and FY83, there is insufficient program funding to do the Senior Trend development tasks currently directed in the PMD....
>
> The contractor cost growth is real. It is a bottoms up estimate which we have reviewed and have, reluctantly, found to be justifiable; with the possible exception of the level of manning required to support flight test....
>
> We have directed ADP to stop work on the fabrication and testing of an alternate tailpipe design, weapon bay tank group provisions, high AOA [angle of attack] provisions, and the threat flight planning document. We also directed them to descope the flight test support manning to their March 81 level.... I took these actions to live within the [FY82 appropriation], but I would like to point out that these actions are inconsistent with the program priorities paragraph of the PMD....[207]

The chief causes for the cost increase were tail redesign activities because of the directional stability problem, fuel cell problems, weapon bay work, continuing tailpipe problems, and rephasing of the flight test program due to the delays caused by all of these. Many of the cost increases were expected to have an impact on FY83 as well as FY82 budgets. Also at this time, F404 engine costs increased because of manufacturing changes that were directed following an F404 engine failure that resulted in the loss of an F/A-18.[30]

This issue was resolved through a combination of reprogramming from production to research, development, test, and evaluation within Senior Trend, adding FY82 money to Senior Trend by reprogramming within the overall low observables programs budget, making selected reductions in the scope of the test program and using Senior Trend management reserve funds. The FY83 Senior Trend appropriation was also increased to help cover these cost increases.[140]

A further funding difficulty loomed on the horizon with the imminent termination of two programs: the L1011 airliner, Lockheed's one venture into the large commercial jetliner arena; and another low observables program (the same one that Senior Trend had "borrowed" money from to help resolve the FY82 funding issue). "A major and critical factor within the funding requirements...is the assumption that [the other program] will be testing...as we have always planned. In the event that [the other program] is canceled, Senior Trend will have to assume the major part" of various costs, including the airlift operations, which had up to that time been shared between the two programs.[206] A handwritten postscript adds, "L1011 [termination] will really impact [Lockheed overhead in] FY83 and on." Both the industry base and the military test facility structure require a certain amount of funding to maintain; the fewer programs there are to share this burden, the more pronounced it will become.

Test Management Issues

The unorthodox management structure and high degree of concurrency in the Senior Trend program created some difficulty in defining roles within the flight test organization and the level of system maturity that was to be expected. The Skunk Works "old hands" initially were bitterly opposed to the whole concept of a Joint Test Force; they felt that the contractor should perform the initial testing, ensure that the aircraft performed as it should, and only then hand it over to the U.S. Air Force.[34] Nevertheless, once the program started, relationships between Lockheed and Air Force personnel were exemplary. Within the U.S. Air Force, friction existed among test site personnel, program personnel, and Tactical Air Command, over who actually controlled the testing. As noted earlier, Tactical Air Command was responsible for operational test and evaluation (OT&E). However, the designated responsible test organization was led by Col Pete Winters, commander of the Air Force Flight Test Center detachment at the test site. Col Winters asserted that it was his duty to test the aircraft against its intended mission and report results. As difficulties emerged, both the responsible test organization and the Combined Test Force* expressed serious concerns over the program to senior DoD officials. These included criticisms of the high degree of concurrency in the program and of the increased production rate that was being contemplated at the time (late 1982). As noted in a message from Lt Col Marks (RDPJ) to Col Scofield at the SPO:

> The briefing presented by the Senior Trend Combined Test Force to Adm Linder [director of testing and evaluation for DoD] and myself on Sept. 14 was the most pessimistic briefing I have heard on the progress of the flight test program. Several development areas were identified as being in trouble with no indication of potential solutions. Specifically, the IRADS screens and the alpha-beta probes were indicated as problem areas that may not have technical solutions at all. If in fact the risks present in the flight test program are as high as indicated, we need to reconsider the merit of accelerating the production program via the 1982 reprogramming....

---

*As the program progressed and became more oriented toward IOT&E, the Joint Test Force (JTF) was renamed Combined Test Force (CTF), reflecting Tactical Air Command's increasing role and increased responsibility for OT&E.[190] Both terms, CTF and JTF, appear to have been used during late 1982.

> Apparently, there is a difference of opinion between the Combined Test Force and Program Office on the potential of the contractor to resolve the problems identified in the test program. To respond to this issue, I am requesting your assessment of the progress of the flight test program.
>
> I am concerned not only by the negative comments made in the briefing, but to whom the comments were made. My intent is not to keep problems out of view, but I would like to be prepared for any revelations. I knew of problems that could delay the progress of achieving an operational capability, but not those that would prevent the aircraft from having a real combat capability. This has been the message I have tried to communicate to the people in Washington. If this optimism is unjustified, let me know.[143]

Two days later, Col Paul Kaminski (RDPJ) followed up with a communication to Col Winters:

> It has been brought to my attention that in the sessions during Adm Linder's visit, there appeared to be a considerable degree of pessimism voiced by you concerning the Senior Trend program.... In addition, there appeared to be some premature judgments made in the JTF [Joint Task Force] briefing regarding the ability of the contractor to design and develop fixes to some problem areas, specifically the options available in the event the changes that are being worked are not successful, e.g., tail hook, IRADS.
>
> I'm concerned in an overall context about making early judgments and passing them on to high level OSD [Office of the Secretary of Defense] staffs before we in the program have reviewed results and reached some common agreement. All of us recognize that there are development issues that are yet to be resolved, but they are being worked. To date, I have been led to believe that hurdles in the program are solvable. If you have results that indicate otherwise, I would like to know and we then need to meet with the program office to follow-up.
>
> In summary, I believe that judgments on corrective actions/redesign options clearly should be worked hand-in-glove with the Program Office before "doom and gloom" bottom lines are incorporated in briefings or discussions with seniors, such as Adm Linder. On the surface, there appears to be a disconnect between your statements and what the SPO says. If there is a void between the JTF and the SPO, then now is the time to get together and sort things out.[112]

A coordinated reply from the SPO's Col Scofield and the Combined Test Force's Lt Col Roger Moseley quickly reassured the RDPJ office:

> The briefing that was presented to you by the CTF [Combined Test Force] was presented in the context of being able to meet near term mile-

> stones not in the context of an unlimited time period. We have structured all detailed test planning, which you saw, to meet the four major milestones that culminate in an initial operational capability on Dec. 31, 1982. If we were to change the milestone dates, the CTF's assessment of the test status would also change.
>
> There is no difference of opinion between the CTF and the Program Office that we know of on the potential of the contractor to resolve the problems currently identified in the test program. The SPO, CTF and contractor meet every other Monday to discuss in great detail the scheduled test program and the actions that ADP [the contractor] must take so the CTF can execute the tests on time. The CTF identifies the current problems that they see in being able to conduct flight testing necessary to support the TAC [Tactical Air Command] requirement milestone date. The company then takes the action within the company to get the problem solved and meet the need date. Never has the CTF expressed an opinion in that meeting, or privately to the SPO following any of those meetings, that we had an unresolvable problem. The schedules and the company plans that come out of those meetings, while success oriented in some cases, address each near term problem that the JTF raises.
>
> At the current time, we do not see any problems that would prevent the aircraft from having a real combat capability, as that capability has been consistently stated over the last four years. We do have some problems that will have to be overcome if we are to meet the specification requirements in all areas. However, we do not see any major problems today that cannot be resolved in the first 107 aircraft test months and result in a weapon system that will meet its intended operational use.[209]

The four major milestones, referred to in the quoted material included weapons testing, clearance of the initial operational flight envelope (including adverse weather operation), delivery of the 10th production aircraft, and training of a corresponding number of TAC aircrew.[101, 146] Although the quote content asserted that these milestones could all be met, given enough time, Col Winters did not completely agree. His comments to the SPO provide insight into the issues associated with concurrent test management, as well as a good description of the status of the aircraft at that time:

> Let me assure you that I have made no judgments, but I have expressed some concern that the ATA program is too optimistic and that there may be some problems which, if not solved, will prevent the airplane from doing its primary mission. On this issue, I do disagree with your message.... It is the duty of the RTO [responsible test organization] to test aircraft against its mission and to accurately report the results. Because of the unusual management structure of the Senior Trend program, the normal checks and balances are not present. This has been apparent in testing as well as production. Since AFTEC [the recently

formed Air Force Test and Evaluation Center, the Air Force's independent test organization] is not on board to independently report on operational suitability, this duty must fall on the RTO's for DT&E [development, test, and evaluation] and IOT&E [initial operational test and evaluation].

I see the primary mission of the ATA as night interdiction with primary weapons delivery using FLIR/DLIR system. Terminal weather must be planned to be without visible moisture but in a European scenario the aircraft must very often penetrate weather with probable icing [enroute to the target, even if weather at the target is clear]....I am concerned that some of the problems we have seen and that are anticipated may prevent the ATA from doing this mission with the survivability desired....

I will briefly list some of my concerns. In some cases these are problems for which solutions have been proposed. In some cases these solutions are insufficient. In others the solutions are simply unproven.

A. Structures: The aircraft continues to have gear problems; i.e., actuator locking rods, unsafe indications, murphy on the nose wheel, and poor QC [quality control]. These are all solvable but indicated poor design and manufacture. The tail hook problems are also solvable but not without structural modifications. Latest tests show that the hook cannot be deployed airborne and skips along the runway when deployed on the ground. An engagement would probably result in the cable striking the aft section of the wings and aircraft. The frangible cover problem should be easily solved. The airplane has two large structural problems. We have reached 100% loads where 80% loads were predicted and cracks have occurred in panels which were predicted to have no stress at all. Both of these indicate that there is poor understanding of stress loads on a faceted airplane. These stress problems are solvable but the solution will probably take a little time. The FLIR/DLIR cavities present a different problem. The latest screens reached their acoustical limits at 400 kn, an air speed which may present serious vulnerability problems. How the FLIR/DLIR will operate in this acoustic environment is unknown. We have never operated an airplane with a hole in the front without a corresponding hole in the rear. The screens may present an insurmountable visibility problem which I will discuss under weather/icing problems.

B. The avionics problems do not appear to be insurmountable at this time. There is, however, no demonstrated FLIR/DLIR capability. I believe the FLIR/DLIR will work if the acoustic environment permits and the screens provide sufficient visibility....

C. All Weather: We have every hope that the inlet de-icing and pitot/α/ß probe heaters will work....The FLIR/DLIR screens presents a greater problem. These screens are not de-iced. They will very probably act much like any screen and ice very rapidly. Electrical de-icing will probably not be acceptable from an RCS or a power required standpoint. Hot air might work but a great amount will be required....the FLIR will not like looking through hot air. The hot air can be turned off in the target area when

> the FLIR is required. One problem remains; the hot air will melt the ice into water which adheres to screens very nicely and is opaque to the FLIR. Bottom line is that I do not think the FLIR/DLIR screen, window, grid or whatever has been completely solved....
>
> I have long maintained that the ATA was pressed into production before it was ready. The program would have been much better designed as a two aircraft prototype, five aircraft pre-production program instead of a concurrent DT&E and production program. With the problems developing if any change to the program is made, we should be slowing down production. No attempt should be made to speed deliveries at this time.
>
> The results of the Senior Trend program will greatly influence future low observable programs, most specifically the ATB [Advanced Technology Bomber, eventually the B-2]. Every care must be taken to test as thoroughly and critically as possible. We cannot let the advocates of this technology quietly influence the course of the Air Force's future weapons.... I strongly suggest that the three of us [Col Winters, Col Scofield, and Col Kaminski] get together in the near future to discuss all the problems and reach an agreement on the status of the program.[255]

In fact, probably due to a combination of production rate versus cost considerations, and the test considerations discussed, the production rate was never accelerated beyond eight aircraft per year. Furthermore, initial operational capability demonstration date did slip more than a year from the originally planned date of March 1982. However, this fact did not result in significant loss of support for the program or any other serious disruptions. Effort was focused on solving each problem as quickly and effectively as possible, without delaying the whole program.

As noted earlier, some inevitable problems had been recognized from the start, and the risks of concurrency had been understood and accepted. To accommodate the risks and focus effort on the most essential tasks, the requirements for a limited initial operational capability had been very specifically defined ahead of time. According to then-Col James Allen, commander of the 4450th Tactical Group:

> That definition, although very limited in actual capability, drove the entire program. It determined what needed to be fixed in production, what required intense engineering work, and what was tested when. Bottom line was that if a system or test was not needed for the limited IOC [initial operational capability] it was delayed. Not all agreed with this approach...but that was the way we did business and it worked.[7]

SUMMARY

That the various challenges were overcome was apparent from a briefing titled, "Senior Trend Test Program from Concept to Initial Operational Capability:

Table 8 Overall performance on F-117A test program challenges[179]

Challenge	Rating
Aircraft	
Radar cross section	Excellent
Two-dimensional exhaust nozzle	Good
Aircraft flying qualities	Good
Aircraft performance goals	Good
All weather operations (ice)	Good
Management	
Security	Outstanding
Flight test organization	Outstanding
Flight test support facilities	Excellent
Remote final assembly and flight test	Good
Concurrency	Excellent

Planned vs Actual"[179] that was presented by Lt Col Moseley to four-star audiences in December 1983, shortly after limited initial operational capability was achieved. In this presentation, challenges associated with the test program were identified and ratings for how well each challenge was dealt with were assigned as shown in Tables 8 and 9.

Specific shortcomings were that contractor representation did not include sufficient technical depth to deal with unanticipated avionics issues. Many irksome airframe problems (for example, battery acid spillage, wiring problems, etc.) were unrelated to the F-117A's unique low observable design, and nearly all schedules were not realistic. However, the CTF found the contractor flight test personnel "honest to a fault,"[179] quite possibly attributable to their Skunk Works training and experience working for Kelly Johnson, who had a reputation as a tough but honest boss who asked a lot of questions and liked to hear direct answers.

The CTF's overall conclusions were that the original schedule was overly optimistic ("you can't avoid problems"), but that concurrency can be handled

Table 9 Contractor performance in the flight test program[179]

Challenge	Rating
New technology	Excellent
Airframe (general)	Fair
Management	Good

as long as some flexibility is maintained. Final recommendations were to maintain better on-site support for avionics integration, ideally with a full fidelity avionics simulator and a flying testbed; plan for downtime due to new technology; provide for update modifications on concurrent programs; and *Do this again!* As Lt Col Moseley noted, "Despite unforeseen and serious challenges a small team was able to develop a radically new and important weapon system in record time, safely, and without compromising the original objective."[179]

Chapter 5

Production Planning and Implementation

Initial Plan

The original plan at the time of the full-scale development go-ahead in November 1978 was to produce 20 Advanced Tactical Aircraft (ATA) in two lots of 10 each. These were referred to as Lot II and Lot III (the five full-scale development aircraft were Lot I) and they were to be funded in FY80 and FY81 respectively. Total production costs were initially estimated at $665 million in FY79 dollars, or $33 million per aircraft.[90] Two of the five full-scale development aircraft were to be upgraded to production configuration for an eventual total of 22 mission capable aircraft. The planned force structure was one squadron of 18 PAA (primary aircraft authorized).[90, *]

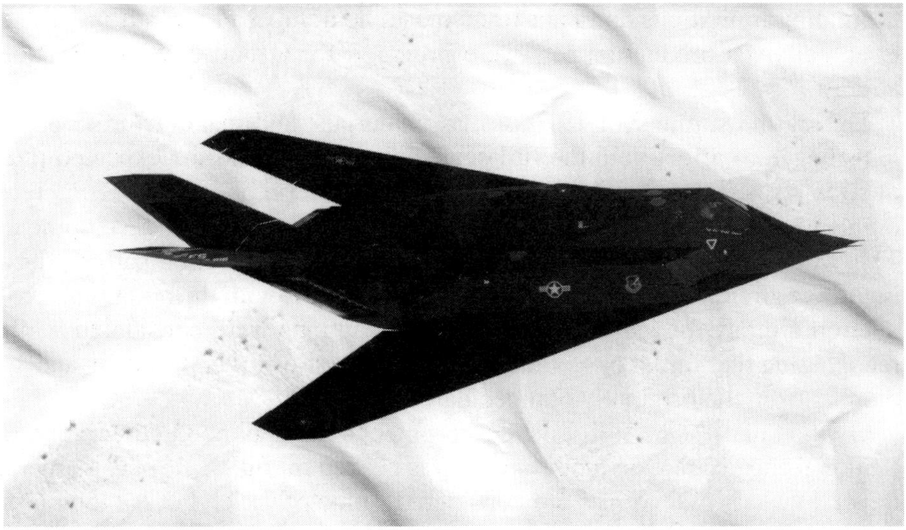

* PAA denotes the nominal operational strength of an Air Force flying unit; i.e., the number of aircraft that can be assumed to be on the flight line at any given time to accomplish the assigned mission of the unit. PAA also forms the basis for allocation of necessary operations and support manpower, support equipment, and flying hours to the unit.

Deliveries would begin in April 1981 and continue at the rate of one per month through December 1982. The first production lot of 10 aircraft would be delivered before March 1982 to support the initial operational capability demonstration scheduled at that time.[84, 240] Because the entire first production lot was to be delivered prior to completing the full-scale development program, a provision was included in the first lot production contract. It required Lockheed to provide modification kits to correct all problems found during initial full-scale development flight testing.[181]

To protect the planned schedule, long-lead ordering for 20 aircraft and 60 engines was authorized in November or December 1979.[240,*] Full production approval was anticipated in February 1980, at which time the cost plus fixed fee long-lead contract would be incorporated into the respective aircraft and engine contracts, both expected to be cost plus incentive fee.[224, 240] However, Gen Alton Slay, who was by this time commander of the Air Force Systems Command, insisted on a fixed price contract.[128] Lockheed finally agreed to a fixed price incentive contract in March 1980, and contract negotiations for the first production lot were completed in late 1980.[217]

Changes and Decision Process in Production Planning

Even before those involved completed the negotiations for the first production lot, the procurement plan changed, first to stretch out deliveries due to funding limitations and later to increase the number of aircraft procured. Eventually, 10 production lots were procured for a total of 59 production aircraft.

During the summer of 1979 design, materials, and manufacturing problems began to affect both the full-scale development and the expected production costs. To "live within" then-current near-term budgets, the SPO recommended transferring money from production to full-scale development. This would leave the entire funding shortfall in the production budget, forcing fewer aircraft to be procured in FY80 and FY81, with the remaining production aircraft funded in FY82.[155] Various options were considered while recognizing that initial operational capability would slip between six and 12 months from the originally planned date of March 1982.

By the fall of 1980, full-scale development costs had mounted further and Senior Trend production money again was needed for full-scale development. Several "what-if" options were considered, some of which left the total pro-

*Some sources report that production was approved in October 1978. Clearly, there was an intent to buy 20 production ATA at that time. However, the first production contract was not in effect until late 1979, as stated above, and was for long-lead items only. Some confusion may have arisen from the fact that the first production lot was "Lot 2." The Lot 1 contract was not for production aircraft, but for the five full-scale development aircraft.

duction run short of the originally planned 20 aircraft. The net effect was that the FY82 procurement was reduced and the production run was stretched into FY83.[22, 156] The major procurement profiles considered, from program initiation through early 1981, are illustrated in Table 10.

A recurring issue was how to remain within total near-term Senior Trend funding, which included funds for research, development, test and evaluation plus production. Requesting changes to current major budget documents or even changing previously submitted budget requests risked attracting unwanted attention to the program. To add money to the total Senior Trend budget without requesting any such changes or obtaining special outside approval, appears to have required a lead time of two to three years. The approach of transferring money from production to research, development, test and evaluation within Senior Trend was therefore used on several occasions, causing production orders to repeatedly slip to the right.

Meanwhile, starting in mid-1980 the U.S. Air Force (with strong encouragement from OSD and those members of Congress who had been briefed into the program) recommended procurement of roughly a full wing of three squadrons of F-117As, rather than just one squadron.[83, 149] Not surprisingly, Lockheed had recommended a larger procurement even earlier: a Lockheed budget and schedule estimate of February 1978 had addressed a possible procurement of 100 aircraft beyond the initial 20.[125] The U.S. Air Force had never completely dismissed this idea and had included the additional buy in budget briefings.[90] The modest size of the initial procurement reflected a philosophy of starting out small, which was appropriate for such a revolutionary new type of system developed under a highly concurrent program.

However, contrary to what seems a widely held perception, a larger quantity was completely consistent with the mission of the aircraft as envisioned very early in the program. Although the ATA was seen by some as a limited-use weapon for special covert operations, others recognized its value in the overall Air Force structure in the event of a full-scale war. The major initial impetus for its development came from experiences with air campaigns in

Table 10 Production profiles considered, 1979 through early 1981

Date and source of estimate	Planned buys by fiscal year funded									
	80	81	82	83	84	85	86	87	88	Total
Original[125, 240]	10	10								20
July 1980[241]	7	8	5							20
Oct. 1980[156]	7	7	0							14
Nov. 1980[157]	7	7	4							18
Early 1981[92, 139]	7	7	4	2						20

Vietnam and from the Middle East war of 1973, which highlighted the need for strike aircraft with greatly improved survivability. Accordingly, since October 1979 the Senior Trend Acquisition Plan had stated:

> The ATA will be employed in offensive operations and in a manner which contributes to increased effectiveness of other friendly weapon systems.... The aircraft will be operated to capitalize on the aircraft's low observable signatures to confuse, disrupt and destroy the enemy's war making capability....The principal targets will be command, control communications centers, air defense facilities, airfields, logistics choke points, and other targets of high military value.[240]

These statements show a clear understanding that the ATA would be a valuable asset in its own right, as well as an important "force multiplier," allowing all friendly aircraft greater freedom to operate and perform their missions more effectively. Furthermore, more than one squadron would be needed under several possible scenarios, for example, in the event of a war in Central Europe.[78, 130]

OSD concurred with U.S. Air Force recommendations but was unable to immediately fund the increase.[83] This raised the issue of production line continuity. Under the then-current plan, production of the original 20 aircraft would end approximately one year before any follow-on production could be funded, therefore, creating a gap in production. Workers with security clearances and experience on the F-117A production line would move on to other jobs, and restaffing would become necessary once further procurement occurred. New workers would require training and security clearance processing time. Significant delays and cost increases would result.

To avoid such a production line break, program managers considered two main options. One was to stretch out production of the already-funded aircraft, avoiding a shutdown but still producing an inefficient "dip" in the production rate. The other option was to bridge the gap by adding aircraft to the FY82 procurement. This would allow a smooth transition into an increased procurement quantity. A final commitment would be made after full-scale development flight testing had progressed to the point of validating radar cross section (RCS) and avionics performance.[22, 92] Rate production studies conducted by Lockheed in conjunction with the System Program Office (SPO) indicated that a production rate of eight to 12 aircraft per year would yield the lowest unit flyaway cost. Lower rates were less efficient, but higher rates would incur additional "tooling-up" costs.[157] Based on these studies, the FY82 procurement increased from the previously budgeted four aircraft (as shown earlier in Table 10), up to a total of eight.

This brought the total planned Senior Trend procurement to an interim level of 24 production units. In addition to the production rate and continuity considerations already mentioned, this would also provide "at least a 10-year force life for the originally planned 18 PAA [primary aircraft authorized] squadron."[83, 92] Presumably there was an understanding that 22 aircraft (20 production plus two upgraded full-scale development aircraft) could not realistically support an 18 aircraft operational squadron and still allow for training, maintenance downtime, and attrition. In April 1981, officials determined that another Air Force program had an excess of funding in FY82 that was sufficient to cover the four added aircraft. An Air Force briefing on FY82 Budget Issues, dated March 1981, recommended applying this money to Senior Trend.[92]

Meanwhile, in anticipation of eventually supporting a *substantially* increased Senior Trend production program, contractors began to place necessary long-lead material orders and develop implementation plans, although as of March 1981, no formal action had been taken to increase procurement. The Air Force budget estimate submission still only provided for four aircraft in FY82 and a final lot of two in FY83.[158] The contractors were becoming concerned about their long-lead investments. The following letter, sent on March 4, 1981 from R.K. Bertelson of General Electric's Aircraft Engine Division to Col Don Merkl, program director at the SPO, illustrates the industry position very well:

> As we have discussed from time to time, we are protecting, at GE risk, long-lead material for 30 1983 engines starting 3/83, which would protect your aircraft production line continuation.... Our current engine lead time is 39 months due to scarcity of critical materials such as titanium, and we are now at 24 months for March 1983 engines. The financial risk to protect lead time up to this point has not been excessive, primarily order placement to hold delivery position with the material vendors. We are rapidly approaching the time when substantially increased investment must be made in the form of material processing and machining.
>
> We intend to continue to protect your follow-on options since we continue to believe that there is a good chance that the government will decide that the program is vital to overall defense enhancement....
>
> Since the FY82 white world budget decisions have apparently been made, we are wondering if the decision to extend our program has also been made...the accelerating investment in material processing/machining as well as overall capacity planning considerations make an early decision extremely desirable. I would also expect that other critical aircraft system hardware may be facing the same inside-lead-time situation.[28]

Merkl immediately notified Col Ken Staten at RDPJ (the low observables/ special projects office at Air Force Headquarters in the Pentagon) of the problem. The immediate issue was how to fund long-lead items for the extra four

FY82 aircraft. Additional funding would then have to follow rapidly if further production (beyond 24) were to be accomplished without serious disruptions to the aircraft and engine production rates. The reply from RDPJ stated:

> Our first opportunity for changing appropriated dollars is FY82.... FY81 money from [another program] can *not* be spent on Trend without approval of Congress which we don't want to ask for. So, Trend must find ways to live within FY81 and prior up to 1 Oct 81 and "get well" starting in FY82... The [funding for the four extra aircraft in FY82] must be in the FY82 appropriation.[219]

Nevertheless, some additional FY81 money was needed:

> Buying additional aircraft and engines in FY82 requires FY81 funding. We have talked to both GE and ADP. While it is not certain, we believe the contractors will 'cover' commitments so long as the Air Force covers expenditures. As such, we need additional FY81 funding....[162]

The problem was elevated to a higher level. Maj Gen Robert D. Russ (director of Operational Requirements DCS RD&A, U.S. Air Force Headquarters), Gen Mathis (U.S. Air Force Vice Chief of Staff), and Gen Lew Allen (Air Force Chief of Staff) decided after discussions in May to go ahead with a transfer of FY81 funds to cover long-lead expenditures.[201] This would be treated as a reprogramming to be verbally approved by each appropriate Congressional committee. The remaining FY82 production funding for the additional four aircraft was added to the FY82 baseline, which had not yet been provided to Congress.[83]

With a plan in place to fund the first four additional aircraft, the RDPJ office requested help in May from the SPO to develop funding profiles for two alternatives: 24 aircraft ("the recently approved program") and 89 aircraft. "There is presently no intent to implement either of the above plans. These plans are only two of several other procurement/production schedules being discussed as alternatives."[221] The two procurement profiles under consideration at that time are shown in Table 11.

As noted in the previous chapter, the first F-117A full-scale development flight took place three days later, on June 18. This undoubtedly provided

Table 11 Production alternatives, mid-1981

Date and source of estimate	Planned buys by fiscal year funded									
	80	81	82	83	84	85	86	87	88	Total
May 1981[221, 242]	7	7	8	2						24
May 1981[221]	7	7	8	12	18	18	19			89

Table 12 Procurement plan according to OSD FY83 budget position

Date and source of estimate	Planned buys by fiscal year funded (procurement prior to FY83 shown in parentheses)									
	(80)	(81)	(82)	83	84	85	86	87	88	Total
Dec. 1981[139]	(7)	(7)	(8)	2	12	18	18	17	—	89

extra confidence for planners to continue to advance the possibility of an increased production run. The "verbal approval" needed from Congress to transfer FY81 long-lead funding for the initial, interim increase (from 20 to 24) was obtained by then-Col Paul Kaminski, the new director of the low observables office at RDPJ, during July 1981 following meetings with key staff members on the House Appropriations Committee, the Senate Appropriations Committee, the House Armed Services Committee, and the Senate Armed Services Committee.[107–109, 222]

By December 1981, the "official" plan was to procure 89 aircraft: "Effective 9 December 1981, the OSD FY83 budget position for Senior Trend to be recommended to the President is as follows [see Table 12]. Direction will be formalized in revised Senior Trend PMD [program management directive]."[139]

The decision to increase procurement beyond the original 20 aircraft was accompanied by a definite decision to expand the scope of the program by adding several activities. These had not been considered justified for only 20 aircraft. They included a full-scale fatigue test airframe, a Weapon System Trainer (simulator*), and several changes to the Senior Trend maintenance concept.[87, 128] The original plan had been for almost entirely two-level maintenance.† A set of deployable avionics vans constituted the only intermediate level maintenance capability in the original plan. With the increased number of aircraft, it became more important to perform basic repairs on the base; therefore some base-level capabilities such as wheel/tire buildup were incorporated in the program. A second set of the avionics vans was also added to

*There was no provision in the original program for a training simulator. As noted in the previous chapter, there was a flying qualities simulator at Lockheed's Rye Canyon research facility. This was used for development and flight test work and would not have been sufficiently representative of a production F-117A to be suitable for operational pilot training.

†Air Force maintenance is normally broken down into three categories: Organic (unit-level), Intermediate (base-level), and Depot. Organic (O-level, or unit level) maintenance consists of remove/replace operations, which can be performed on the flight line. Intermediate (I-level, or base level) maintenance consists of troubleshooting and repair on base, whether at home or deployed. Depot maintenance consists of major repairs. O-level and I-level maintenance are normally performed by the Air Force. Depot maintenance may be performed either by the Air Force or the contractor. "2-level maintenance" refers to the elimination of I-level maintenance.

allow two squadrons to deploy independently to different theaters. In conjunction with the Air Force assuming more of the maintenance responsibility, some technical orders (for example, manuals for operating and maintaining a system) were produced, although not as many as would normally be the case.[194] This would become an issue much later, because those technical orders that the Air Force did not purchase up front were in fact never produced at all, making the Air Force almost permanently dependent on certain types of contractor maintenance.

With the extended production run, a substantial number of aircraft would be produced after the completion of full-scale development. The later aircraft would incorporate design improvements and modifications resulting from test and early operational experience. A new model, the F-117B, was also contemplated at this time. The F-117B, if developed, would incorporate further technology improvements and a restructured program ("gray" security, operation at a second base, and a revised production line). The technology improvements would include the installation of a low probability of intercept radar, a global positioning system unit for enhanced navigation, and provisions for carriage and delivery of the high speed anti-radiation missile (HARM).[111, 208]

Although the new plan called for a production rate of 18 aircraft per year for the last three years of the production run, the existing facility could only support a rate of 12 per year. Estimates of the time and cost associated with "tooling up" for a higher rate are shown in Tables 13 and 14.

Funding for long-lead items and tooling would have to be provided 12 months in advance of starting fabrication of additional production aircraft. The fabrication and assembly time was 19 months.[94] These estimates indicated that 18 aircraft per year was an attainable, cost-effective figure.

However, Lockheed ADP was having trouble keeping up with the original production schedule. This was due in part to the difficulty of obtaining sufficient manpower and materials and in part to the challenges associated with producing the F-117A. When the program was small, this had not been considered much of a problem; the program had initially been conducted on a "what-you-see-is-what-you-get" basis, with the Air Force allowing the contractor a relatively free hand to accomplish development and production, taking extra time as needed to overcome problems. However, as high-level Air Force, DoD, and Congressional interest in the program picked up, pressure

Table 13 Predicted F-117A production rate buildup capability[94]

Year of buildup	1	2	3	4	5	6
Max. rate, per year	18	38	42	48	60	80

Table 14 Additional tooling-up costs[94]
(relative to a basic F-117A production rate of 12 aircraft per year)

Rate, per year	18	24	48
Increased tooling costs	$15M	$60M	$80M

increased to keep to the originally contracted schedules.[35, 194] Accordingly, on July 21, 1981, Col Don Merkl sent the following memo to Norm Nelson, Lockheed's project manager, and to Lt Col Mike Marks at RDPJ:

> In July 1980, the Air Force and ADP jointly worked to resolve a serious manpower shortage problem that was significantly impacting the Senior Trend delivery schedule. At that time, commitments were made by very senior Lockheed officials pledging priority support to this important, DX-rated program [DX stood for Defense Expedite, a designation for extremely high priority programs].
>
> Subsequent to the "resolution" of the manpower problem last summer, the Senior Trend Lot II production contract was negotiated and signed with the delivery schedule shown in Attachment I [this schedule called for deliveries of the seven aircraft at the rate of one per month from October 1981 through April 1982]. In March 1981, ADP indicated that the Lot II delivery schedule was in jeopardy. At that time... [an independent review team] produced a "revised" delivery schedule that was "doable."
>
> At a meeting on July 6, 1981, the Air Force learned that the Senior Trend program was again short of manpower (314 vacant spaces) in several areas.... There was evidence to suggest that the "doable" schedule had slipped.
>
> The fact that Lockheed management has allowed for the second time the manpower shortage problem to impact the Senior Trend delivery schedule is most disturbing. In addition, the lack of timely management action violates a pledge made last year to General Skantze relative to Lockheed company commitment to the Senior Trend program.
>
> It is apparent to me that Lockheed management is not taking timely corrective action in an attempt to regain the Lot II production schedule. This apparent inaction has seriously endangered the performance of the production contract and represents the sort of management that cannot be tolerated for a program so important to the Nation's defense....[170]

Shortly afterward, "based on Air Force unhappiness" with Lockheed ADP program schedule performance and the transition from development to production, the company made major changes in the program management team in early 1982. In February Sherman Mullin, a 21-year Lockheed veteran and experienced program manager, was assigned as vice president and program manager, reporting to Ben Rich. In March, John Sheridan, a veteran project engineering manager, joined the program as deputy program manager and chief engineer."[181]

At the same time, the accelerating test program was also uncovering new problems in late 1981 and early 1982, and funding again had to be reprogrammed from production to research development, test and evaluation. Because of this and other budget pressures, the planned FY84, FY85, and FY86 procurements were reduced during early 1982. The possibility of reducing the total production run from 89 aircraft to 57 aircraft to support two operational squadrons of 18 PAA each was also discussed. There was even an "air staff directed alternative" to stop Senior Trend production at 33, which included only those funded through FY84, at least until the F-117B was ready to be introduced.[208] The main options at this time are shown in Table 15.

In all three plans, the FY86 budget would be the decision point to provide long-lead funding for F-117B, continued F-117A production, or to terminate production altogether. Senior Trend management argued against the air staff directed alternative, emphasizing as they had in the past the need to maintain production line continuity:

> One cost, which is not totally known at this time, is the cost associated with having to hire the work force back nine months to one year ahead of time in order to allow for proper security clearance processing....
>
> A potential impact that is not quantifiable is that Lockheed California may not be able to survive a production line gap on this program. With the phase-out of the L1011 by 1983, the loss of an ongoing production program, even with a minimum rate, could conceivably result in the demise of the Lockheed California company....[208]

Another source of concern was the provision for upgrades and improvements in the event that neither the 89 aircraft buy, nor the F-117B, were funded. On April 8, 1982, Lt Col Mike Marks at RDPJ had reported that:

> The Air Staff Board, during the FY85 budget drills, has maintained an alternative to end Senior Trend production of 57 aircraft (56 aircraft force structure). Although this alternative has not been implemented, there

Table 15 Senior Trend procurement alternatives as of April 1982[208]

	Planned buys by fiscal year funded									
Alternative	80	81	82	83	84	85	86	87	88	Total
Prior plan	7	7	8	2	12	18	18	17		89
"Best case"	7	7	8	2	9	12	12	F-117B⇒		57 F-117A
"Air staff"	7	7	8	2	9	0	0	F-117B?⇒		33 F–117A
Compromise	7	7	8	2	9	<12	<12	F-117B or add'l F-117A		TBD

> is enough support among the general officers on the board to make it happen. The unresolved budget pressures in the out years may make this option more attractive as final budget decisions are being made.... We need to have a backup plan for incorporation of improvements.
>
> I recommend that all planning for follow-on improvements to the Senior Trend aircraft include the retrofit of existing operational aircraft. This planning should include cost estimates, schedule, and technical feasibility of aircraft modifications for a force structure consisting of 56 aircraft based at Tonopah. The planning for a force structure of 56 aircraft would have equal importance as the planning for 89 aircraft.[141]

The RDPJ staff's very important foresight helped ensure that planning would address modifications and upgrades to the entire F-117 fleet and not just to those at the "tail end" of the production run. This proved to be a wise precaution. At a meeting on April 26, 1982:

> ... the Air Force position was firmed to procure 56 Senior Trend aircraft to support two 18 PAA squadrons at a rate of eight per year. Funds were put into the FY84 POM, to procure eight aircraft and to provide advanced [long lead funding] for eight more to be fully funded in FY85.... At this production rate, the FY87 POM would be key decision point for the procurement of a third squadron (full wing)...it was also concluded that a full wing (89 aircraft) can be based at Tonopah under our current security concept.[153]

The last point was apparently a matter of some question, as Col Paul Kaminski noted in a May 21, 1982, memo that two was probably the maximum number of squadrons that could be based covertly at Tonopah without significant new construction.[111] This was probably a more realistic assessment and was one of the reasons that a restructured "gray-security" program was envisioned for the F-117B, with operations at a second, noncovert base.

Meanwhile, significant events had taken place at the test site. The first round of RCS tests were completed in January and February 1982. By the spring of 1982, all five full-scale development aircraft had been delivered and all but one had flown (Aircraft 783 was still undergoing RCS preparation).[211]

The first production F-117A (number 785, picking up the tail numbers where the full-scale development aircraft had left off) had also arrived at the test site. It crashed while attempting its first takeoff on April 20, 1982, prior to its formal acceptance by the Air Force. The pitch and yaw rate gyro input to the flight control computer had been cross-connected. When the aircraft rotated (pitched up) for takeoff, this motion therefore appeared to the computer as an uncommanded yaw departure. The computer applied "corrective" rudder, causing a *real* yaw departure, which the computer perceived as pitch. The computer "corrected" again with full-up elevons, and the combination of abrupt pitch and yaw input apparently snap rolled the aircraft. It landed on its back and was a total loss. Lockheed pilot Bob Ridenauer survived but was seriously injured with two broken legs.[96] Testing of the control system on the ground had not revealed any problem. Without actually moving the aircraft, there had been no way to tell that the gyros were incorrectly wired.

According to Sherman Mullin, the Skunk Works management was shocked by this accident and quickly decided "to have an independent Lockheed audit of the program and share the results with the U.S. Air Force. W. Paul Frech, a Lockheed vice president and an authority on aircraft production, was assigned to lead the audit. Col Dick Scofield, the Air Force program manager for the F-117A, cleared Frech and 15 other Lockheed veterans into the program to conduct the six-week audit. The result in June 1982 was the Frech Report, which contained approximately 100 corrective actions to improve quality and schedule performance on the production program. All of these were implemented...."[181] The corrective actions included, among many others[96] redesigned gyro connectors with physical differences to prevent misinstallation, revised flight control system checkout procedures, and development of a configuration audit and reinspection of all completed aircraft.

Col Scofield, Lt Col Jim Bottomley, SPO manufacturing director, and others were kept informed on the progress of the corrective action program. As noted by Sherm Mullen, more senior Air Force leadership attention was also now involved; "Lt Gen Tom McMullen, ASD commander, came to LADP Burbank in August and again in November 1982 for in-depth program reviews."[181]

The acceptance of subsequent aircraft was slowed by the additional procedures, and this compounded delays in production deliveries that had already been a source of frustration. Following the accident, from April through August 1982, no production F-117As were delivered. Lt Col Jerry W. Baber, business and financial manager at the SPO and formerly one of the original five RDPJ staff, at this time sent the following message to Lockheed ADP, with an informational copy to RDPJ:

> As of Aug. 23, 1982, ADP is delinquent in the delivery of Senior Trend Aircraft 786. Therefore, in accordance with Contract Modification P00018, paragraph 3, progress payments are hereby suspended on the subject contract. This suspension will remain in effect until said aircraft is delivered.[24]

A handwritten note, presumably on the RDPJ copy only, stated:

> Mike [Marks] and John [Madia] – The above referenced contract modification was a bilateral change, negotiated and agreed to by the contractor. Just so you don't get the idea that I am being arbitrary and capricious.[24,] *

---

*An interesting point to mention is that Lt Col Jerry Baber presently holds the position of director of contracts at the Lockheed Martin Skunk Works.

The next production aircraft to be delivered was actually number 787, accepted by the U.S. Air Force on August 23, 1982, the same day as the memo quoted above. To have been accepted on August 23, it must have already been at the test site for some time to accomplish the necessary pre-acceptance testing. Aircraft 786 followed soon after and was accepted on September 2, 1982.[179] "By late 1982 the production program had stabilized and a new production aircraft was being delivered to the 4450th Group at Tonopah every six weeks," recalls Mullin.[181]

By this time the procurement plan had essentially converged on a schedule for the production of 57 aircraft that is very similar to what was actually executed. Production would be maintained at the more realistic rate of approximately eight per year rather than 18 per year as contemplated in some of the earlier plans. The history of planned F-117A procurement is summarized in Table 16.

Many other "what ifs" options also existed. Only the "major" changes are shown in Table 16. Even though the plan changed little between 1982 and the actual completion of the production run, each procurement lot still underwent a decision and review process before funding.

As noted earlier, the program also initially planned for upgrading two of the full-scale development aircraft to operational status while using the remaining three for continued testing. This plan was dropped in early 1982; a memo from Capt Ferry in the SPO to Lt Col Marks recognized that the full-scale development aircraft "most probably...will be too 'tired' to refurb when testing is complete."[62] The final outcome, which certainly would not have

Table 16 Summary of F-117A procurement profile options[a]

Date and source of estimate	Planned buys by fiscal year funded									
	80	81	82	83	84	85	86	87	88	Total
Original[125, 240]	10	10								20
July 1980[241]	7	8	5							20
Oct. 1980[156]	7	7	0							14
Nov. 1980[157]	7	7	4							18
Early 1981[23, 92]	7	7	4	2						20
May 1981[221, 242]	7	7	8	2						24
May 1981[221]	7	7	8	12	18	18	19			89
Dec. 1981[139]	7	7	8	2	12	18	18	17		89
Nov. 1982[243]	7	7	12[a]	2	8	8	8	5		57
Actual[128, 245]	7	7	12[a]	2	8	8	8	4	3	59

[a] Includes two production lots, Lot 4 (the original FY82 lot of eight aircraft) and Lot 6 ("Supplemental FY82 lot" of four aircraft added by Congressional reprogramming very late in FY82); Lot 5 was the FY83 lot of two aircraft.

been anticipated during the early days of the program, was that the two full-scale development aircraft not used for continued testing are now on public display. Number 780 is mounted on a pylon at Nellis Air Force Base, Nevada, and F-117A number 781 is in the Air Force Museum at Wright–Patterson Air Force Base, near Dayton, Ohio.[175] Another display aircraft, referred to as "785" but actually reconstructed with parts from various F-117As including the static test article, is mounted outside the current Skunk Works location (now the Lockheed Martin Skunk Works, in Palmdale, California.[232,*]

The actual F-117A production by lot number and the number of aircraft delivered by year are summarized in Tables 17 and 18.

The important point is that the budgeting maneuvers were executed successfully enough to allow the production line to physically turn out airplanes at an efficient, steady rate of eight per year for the first five years of production, followed by a gradual tapering off to the end of the production run. Providing long-lead funding for airframes and engines was one of the most important elements in this process.

As the Senior Trend program progressed, and more people became aware of its existence and of the difficulties being encountered, funding for the program came under increasing attack. Several DoD program budget decisions were issued in late 1982, which reflected a lack of understanding or an unwillingness to accommodate the unusual aspects of the Senior Trend program. One example, undated but apparently late November 1982, recommended against increasing the production rate and also charged that the F-117A had "no mission" and "no force structure."[184] The program budget decision presented an alternative budget that cut out-year research, development, test and evaluation funding because these activities should phase out as the program moved into production and operations. The budget also delayed operational support funding to match the slip in initial operational capability and eliminated all production funding from FY84 and on, pending better definition of the mission and the force structure. In fact, these recommendations were based on certain misconceptions, and on a lack of understanding of how the Senior Trend program was conducted.

The strategy of fielding the system as rapidly as possible, and following up with upgrades and "fixes" on a continuing basis, required that testing activities continue at a higher-than-typical level into the production and operation phases of the program. The mission was well defined and included destroying high-value/high-leverage targets in a high-threat environment, which other aircraft would not be able to penetrate. As noted, the mission need had been identified through recent military experiences in Vietnam and the Middle East, and the definition of the mission requirements was supported by extensive

*Lt Col Jerry Baber, now at the Skunk Works, likes to have his lunch outside under "Aircraft 785."

Table 17 F-117A aircraft deliveries by year[171, 240, 245]

CY	81	82	83	84	85	86	87	88	89	90
Original plan	8	12		(possible follow-on buy of 100)						
Actual deliveries	—	7	8	8	8	8	7	5	4	3

threat analysis. The force structure was also well defined: 56 aircraft would provide for two squadrons of 18 PAA each, plus four backup inventory aircraft, nine training aircraft, and seven advance attrition aircraft.[147, 148]

However, few people outside of the Senior Trend program itself knew about the extensive concept definition and requirements development efforts or appreciated the breakthrough magnitude of the concept and technologies that were embodied in the F-117A and that had been validated by Have Blue. The unconventional development and management approach used to acquire this revolutionary new technology into operational use as quickly as possible aroused suspicions and led to criticism of the program.

PRODUCIBILITY

While the increased F-117A procurement was, as noted earlier, consistent with the intended mission of the aircraft, it did introduce problems related to production methods. Because of the small size of the initially planned production run, less attention had been paid to producibility than would normally be the case. In fact, F-117A production was extremely labor intensive.

Table 18 Actual F-117A production by lot number[128, 240, 245]

Lot no.	No. of aircraft	Contract type	Year funded, FY	Acceptance Dates: Start	Acceptance Dates: Finish
2	7	FPI	80	Aug. 1982[a]	Dec. 1982
3	7	FPI	81	Dec. 1982	Oct. 1983
4	8	FPI	82	Oct. 1983	Sept. 1984
6	4	FPI	82	Nov. 1984	April 1985
5	2	FPI	83	March 1985	June 1985
7	8	FFP	84	July 1985	June 1986
8	8	FFP	85	Aug. 1986	June 1987
9	8	FFP	86	Nov. 1987	Oct. 1988
10	4	FFP	87	Feb. 1989	Nov. 1989
11	3	FFP	88	March 1990	July 1990

[a]The first unit of Lot 2 crashed prior to acceptance; the date shown is for the second unit.

At around Lot 4, the Air Force therefore negotiated with Lockheed ADP for a producibility improvement program.[128] Two phases were proposed, with phase I to begin at the 37th production unit (the first aircraft of Lot 8) and phase II at the 53rd (the first aircraft of Lot 10). Because the production run ended at 59, it is assumed that only phase I was implemented. This phase was projected to cost $21.9 million and yield a savings of $2 million per aircraft.[127] As Col Mike Marks noted, the Air Force authorized this effort "based on a Lockheed commitment to show at least a two to one payoff in reduced production costs. It is generally agreed that the company bettered that objective."[128] The largest components of this program were tooling and coating improvements. Since late 1980, Lockheed had been experimenting with ways to spray on the RAM coating as an alternative to the early technique of gluing it on in sheets. By mid-1985, automated spraying processes were being tested on entire wings.[232] The following statements, prepared in 1985 by the Air Force Special Projects Office at the Pentagon (by then referred to as RDQ-LO; formerly RDPJ), concisely summarize the F-117A producibility situation:

> Initially, an extremely conservative approach was taken in the coating area. Operational experience has shown that more access doors can be added without degradation in the RCS area. Lockheed is moving in this direction along with spray on RAM versus the current glue on coatings. Within one year, the wings and major portions of the F-117A fuselage will be coated with spray-on RAM....

> In the area of manufacturing, Lockheed is currently working with very manpower intensive assembly lines on Senior Trend. This is partially caused by the low rate of production (eight per year) and their historically small production programs (U-2 and SR-71). This has resulted in a perception of a "hobby shop" operation. They have an ongoing effort to improve this deficiency but due to the low total Senior Trend buy (56), they cannot economically update their manufacturing as much as might be desirable in a larger program.[75]

What was done was probably the best that could be done under the circumstances, which was to keep building airplanes at the rate for which the production line was designed. Modest producibility improvements were made as appropriate to the small number of aircraft remaining to be built.

EFFECTS OF DEVELOPMENT CHANGES ON PRODUCTION

The F-117A underwent an unusually large number of design changes during the course of its production run. Some of these are described briefly below in the context of their effects on the production run. The initial tailfin area increase was made in time to incorporate on the first production aircraft. The flutter problem, which precipitated the design of the later composite tail surfaces, was not discovered until 1985, so the production run was almost complete by the time the design of the new fins was completed, and they had to be retrofitted to most aircraft. Structural modifications to the wing root were incorporated at the 24th production aircraft and had to be retrofitted to earlier aircraft.[215] Some tailpipe and nozzle fixes were also made about half-

way through the production run, and therefore required many aircraft to be retrofitted. The various IRADS difficulties (screen issues, acoustics, and FLIR/DLIR alignment) required about one year to fix, but this occurred early enough that it had little impact on the production program.

The unit flyaway cost averaged $42.6 million (then-year dollars) over the entire F-117A production run. The production costs appear to be quite reasonable by modern standards, in spite of the large number of design changes made during the course of the production run. The low production rate of eight per year for most of the production run and the Lockheed Skunk Works procedures for implementing changes in an aggressive and streamlined manner (as discussed in Appendix C), helped minimize production costs associated with midstream design changes.

Perhaps a greater cost effect (although more difficult to quantify) lies in maintaining and supporting the large number of different configurations. At one time there were seven distinct F-117A configurations, comprising different combinations of coating materials or techniques, and aircraft at different stages in the upgrade and retrofit programs.[53] A concerted effort has been made to reduce this number. All F-117As are or soon will be in one standard configuration, at least from a maintenance and support standpoint.

F-117A Production Summary

1) The original plan was for 20 aircraft in two lots of 10 each. However, early concept studies had considered the possibility of a follow-on procurement of an additional 100 aircraft, with a decision point occurring after initial flight and RCS tests.

2) The first long-lead production contract was awarded in late 1979.

3) Funding constraints caused the original program to slip from 10 aircraft each in FY 1980 and FY 1981, to seven, seven, four, and two in FYs 1980 through 1983.

4) The U.S. Air Force and Congress were interested in procuring a larger quantity. The Tactical Air Command concept of operations called for more than 20 aircraft—roughly a full wing.

5) An interim increase from 20 to 24 F-117As was arranged in early 1981 to maintain production continuity. Funding for these four aircraft (reprogramming from another Air Force program) was approved in late 1981, by which time the first flight in the full-scale development program had occurred.

6) By December 1981 the "official" position was that 89 aircraft would be bought, although a final decision was to be made only after the completion of certain avionics and RCS testing. Peak production rate was to be 18 per year.

7) Other significant changes in program scope accompanied the decision to increase the quantity (Air Force maintenance, full scale static test article, training simulator, and so on).

8) During 1982 the procurement was reduced to 57 aircraft produced at a rate of eight per year to keep within available funding, existing production rate capability, and security practices (covert operations at Tonopah).

9) The actual number of F-117As produced was 59 at a peak rate of eight aircraft per year.

10) Throughout the program, short-term funding shortages caused continual slips. Slips also occurred when diverting money from production to research, development, test and evaluation.

11) Nevertheless, budgeting maneuvers were executed successfully enough to allow the production line to physically turn out airplanes at the relatively efficient, steady rate of eight per year. Providing long-lead funding for airframes and engines was one of the most important elements in this process.

12) Even though the procurement plan changed little from late 1982 onward, each fiscal year procurement underwent a separate decision process prior to funding.

Chapter 6

IOC and Beyond

Preparations/Operational Unit

The unit that would eventually operate the F-117A, the 4450th Tactical Group, was formed on October 15, 1979, essentially concurrent with the decision to put the F-117A into production. Col Robert A. Jackson was its first commander.[49] This unit, like the offices that oversaw the F-117A's planning and development, reported to top decision makers through a short and direct line of authority. Specifically, the commander of the 4450th Tactical Group reported directly to the Tactical Air Command director of operations. In contrast, an ordinary wing would report through an Air Division and a "numbered" Air Force (the Ninth Air Force) before reaching the Tactical Air Command director of operations. Although called a group, which is the next level below a wing, the 4450th was essentially a wing. However, it was called a group in order to maintain a low profile.[7]

There were not yet any airplanes to fly at the time the 4450th Tactical Group was established. However, the first members of the operational unit became heavily involved in the development of the F-117. According to Brig

Gen James S. Allen, U.S. Air Force (Ret.), who as a colonel became the 4450th's second commander in May 1982,

> The operators were involved every step of the way once the production decision had been made. The first ops pilot was sent to Burbank [the contractor facility] within six months of that decision.... Members of the operational unit became an essential part of the program team and directly impacted the following areas: operational requirements, functional requirements, concept of employment, hardware selection, cockpit layout, pilot workload, mission planning system, maintenance and support concepts, pilot and maintenance training, and technical publications. Operational unit involvement was such that we had three Mitsubishi MU-2 turboprop executive aircraft assigned, which constantly shuttled people back and forth from Las Vegas to Burbank so that they could work hand in hand with the contractor and SPO.[7]

The F-117A test location was not suitable for an operational base, so a new covert base had to be established. Tonopah Test Range, an area northwest of Nellis Air Force Base which had been used for drop tests of early nuclear weapon "shapes," was chosen as the site for the new base.[229] An extensive military construction project was begun to turn the austere airstrip at Tonopah into a full-fledged base with maintenance facilities, hangars, housing, and so on. Members of the 4450th Tactical Group participated in the planning and development of the base, as well as developing the aircraft.

Because even the existence of the F-117A was classified, strict measures were observed to prevent any chance sightings by outsiders. Although remote, Tonopah could be seen from public land, so pilots flew the F-117A at Tonopah generally only at night. Operational pilots therefore made their first daytime familiarization flights in the F-117A at the test site before joining the flying operations at Tonopah. Because of the tight restrictions on F-117A flying, a surrogate aircraft was needed for training and practice and to provide a cover story for the 4450th's existence.

The 4450th Tactical Group, therefore, operated about 20 Ling-Temco-Vought A-7 aircraft, primarily single-seat A-7Ds but also several two-seat A-7Ks.[7, 190] The A-7D was chosen "because it demanded about the correct amount of pilot workload expected in the F-117, was single seat and therefore would bring all pilots to a common flight training base line, remembering we had pilots who had never flown single seat aircraft outside of solo flight in flight school," according to Gen Allen.[7] For example, some of the F-117A pilots had F-4 or F-111 backgrounds. In addition to providing an excuse for the 4450th's existence and activities (the A-7s were nominally for "avionics testing"), they were also used to maintain pilot currency, particularly in the early stages when very few production F-117As were available. The pilots learned to fly chase on F-117A test and training flights, perform practice covert deployments, and practice any other purpose that could not be accomplished using F-117As, given the tight restrictions imposed on all F-117A operations.

Pilots and ground crews for the 4450th were carefully selected. Gen Allen recently recalled that

> We had an abnormally high number of Fighter Weapons School graduates and below the zone selectees [in other words personnel selected for promotion earlier than normal for a given rank]. For example, in a normal wing there would be 15 to 20 selections to master sergeant each year. However, in the 4450th we would typically have more than 100 selected each year.[7]

Limited Initial Operational Capability

The Senior Trend program was oriented toward fielding a limited operational capability as rapidly as possible. Several aircraft capability limitations were deliberately accepted at initial operational capability to achieve this objective. The term "limited initial operational capability" was therefore used when referring to the F-117A becoming operational with the 4450th Tactical Group. Very specific goals were established preparatory to the full-scale development flight test program, defining what had to be accomplished by limited initial operational capability and what did not.[7, 101] According to Skip Hickey, chief engineer for the F-117A at the SPO,

> The goals covered the areas of flight mechanics, structures, avionics and weapon delivery integration, propulsion, and subsystem operations.... It must be emphasized that the SPO, contractor, and CTF [Combined Test Force] worked together (and most times harmoniously) to provide an F-117A that met those goals and was lethal in the level weapons delivery mode by initial operational capability.[101]

First flight had slipped 11 months from the original plan and initial operational capability by 19 months, so the time from first flight to initial operational capability was eight months longer than planned. The level of effort was also increased (nearly 50% more people), and even so, less flying was accomplished and there were several limitations on the aircraft's capabilities and flight envelope at initial operational capability. Maximum speed, angle of attack, and *g* limits were all lower than planned.[179] However, this did not take away the F-117A's capability to perform its basic mission, as its lethality and survivability did not depend on speed or maneuvering. Weapons testing was entirely focused on achieving a capability for straight and level weapon delivery. Loft and dive modes were never a goal in the full-scale development test program,[7, 101, 179] although the system operational concept of 1983 did include alternative delivery modes as well as air-to-air and anti-shipping capabilities.*

Very ambitious reliability and maintainability goals were also stated in the 1983 Tactical Air Command system operational concept document.[78] The subsequent 1985 and 1987 system operational concept documents were progressively more realistic in their discussion of F-117A system goals as a result of experiences gained during actual operational use of the aircraft.[79, 80]

A Combined Test Force briefing compared planned and actual schedule, level of effort, accomplishments and system capabilities as shown in Table 19.[179] This provides some insight into the challenges encountered in implementing this highly concurrent program. One cause of the delays, relative to the original

---

*According to Lt Col Stan Siefke (who as previously noted was responsible for aircraft performance testing within the CTF)

> ...Although they [alternate weapons delivery modes] were never incorporated into the flight manual, these types of delivery modes had certainly influenced the development of the flight control computer (FLCC). The early flight manual had limitations on how high and low the flight path could get before recovery maneuvers had to be initiated due to the uncertainty at the time about the high angle of attack and limiter performance. Almost all of the rolling maneuvers performed during the angle of attack limiter work were developed by examining the normal roll rates developed during loft and dive maneuvers by typical attack aircraft. Delivery of the B61 [nuclear] weapon dictated the use of a loft maneuver to ensure aircraft survivability, and so maneuvers such as low altitude drogue deliveries and dive/toss deliveries were used to shape the requirements for the handling qualities desired from the FLCC.[212]

Table 19 Test program, planned vs actual (Ref. 179)[a]

Event	Planned at full-scale development go-ahead	Actual	Change
First flight (from Nov. '78 go-ahead)	20 months	31 months	+11 months
IOC (from go-ahead)	40 months	59 months	+19 months
Development funding (up to IOC)	$504.4M	$773.2M	+53%
Production funding (scaled to first 20)	$665.0M	$861.5M	+30%
Production run	20 (special operations)	59 (two tactical squadrons)	Nearly tripled
Up to IOC			
Sorties	850	498	–41%
Flight hours	1350	721	–47%
A/C test months	85	41	–52%
Calendar months	20	28	+40%
Peak personnel	439	620	+41%
At IOC			
Weapons	5 (3 weapons: Mk84, GBU-10, SUU-30; 2 training bombs: BDU- 33, Mk106, also B61 (goal)	3 (2 weapons: Mk84 and GBU-10; 1 training bomb: BDU-33)	
Delivery modes	Level	Level	No change
Max Mach number	0.9	0.8	–0.1 Mach
Max speed	600 kn EAS	470 kn EAS	–130 kn EAS
Max load factor	+6 *g*	+4 *g*	(80% loads)
Max AOA	14 deg	10 deg	–4 deg
Deployable A/C	10	8	–20%
Pilots	10	8	–20%

[a]Some changes made from Ref. 179 to reflect "original" initial operational capability goal of March 1982.

schedule, is that the original schedule was extremely optimistic. According to Lt Col Charles Pinney, a former member of the Combined Test Force, "The program policy was always 100% optimistic scheduling.... The result was constant slips."[190] Nevertheless, the schedule—and cost—of the F-117A full-scale development program are still quite impressive. A key factor that made the program successful was a realistic understanding up-front of the challenges. This understanding resulted in a willingness to accommodate those challenges and a problem-solving, cooperative attitude among all involved.

Program managers and operational users of the system made intelligent compromises to accept those limitations, both planned and unplanned, which

Table 20 4450th Tactical Group's combat readiness ratings at limited initial operational capability[8]

Measured resource area	C-1	C-2	C-3	C-4
Personnel		X		
Training				X
Equipment readiness				X
Equipment supplies on hand	X			

were not critical. This was consistent with the philosophy of preparing the system for use quickly so that experience could be gained in the operation of very low observable aircraft. However, limited initial operational capability was not declared until those issues that were critical (for example, the air data probes and the IRADS screens) were resolved at least to the point of providing a basic mission capability. This was achieved in October 1983. According to the initial operational capability report from the 4450th Tactical Group commander, [8]

> On October 28,1983, the 4450th Tactical Group declares a limited initial operational capability (LIOC) IAW the following criteria:
> a) Possess 10 F-117s configured to limited initial operational capability.
> b) Maintain eight operationally ready F-117s.
> c) Be able to generate
> 1) Eight F-117s in 48 hours.
> 2) Remaining in 72 hours.
> d) Be capable of delivering level
> 1) GBU-10 LGB.
> 2) MK-84 LDGP.
> e) Be capable of deployment worldwide.

The commander's report went on to assign 4450th Tactical Group combat readiness ratings (C-ratings) at limited initial operational capability as shown in Table 20.*

> The overall rating of the 4450 TG [Tactical Group] is C-3; however, it should be clearly understood that the daily rating of our combat capability will fluctuate markedly between C-1 and C-4 due to the small number of aircraft....[8]

The commander's initial operational capability report also listed several factors that were considered to limit the 4450th Tactical Group's capability at limited initial operational capability, in addition to those noted earlier:

*C-1 = Fully Combat Ready; C-2 = Substantially Combat Ready; C-3 = Marginally Combat Ready; C-4 = Not Combat Ready; C-5 = Service Programmed, Not Combat Ready (Source: AFR 55-15, 22 Nov 82).

Mission Planning: The expected wartime daily UTE rate of 2.0 [sorties per aircraft per 24-hour period] in the current SOC [System Operational Concept] is over ambitious at this stage of development of the weapons system.... The reason is that until the computed Automated Planning System (AMPS) is operational (estimate FY 85/86), all mission target planning must be done manually...the time required is six to eight hours dependent upon the threat array...an initial first week wartime rate of six sorties per night for eight deployed aircraft would be realistic.

Aircraft Factors:

Platypus Overheats: The ability to deploy worldwide rests with solving the platypus problem...until we have endurance tested the aircraft, I am not completely confident that we can endure [on a long-range deployment flight].

Weapons Delivery Accuracy: The Dash-five, version 30 software for the aircraft had some significant bombing errors.... Through our insistence and aggressive ADP attention, a new tape, Dash-six, version 34, has been issued, which corrected many of the errors.... Numerous BDU-33 drops by our MR [Mission Rated] pilots show an accuracy of 200–300 ft scattered about the target. We are confident we can deliver on area targets (runways, etc.) at this time.

IRADS: We are experiencing difficulty with the IRADS system...there are severe shortages in both the LRUs [Line Replaceable Units] and SRUs [shop replaceable units]. Our WRSK [war readiness spares kit] is 98% filled, but the 2% not filled appears as critical LRUs and SRUs.... We, with the aid of ADP, will extract spare parts from the production line for deployments.

RCS/RAM Degrade: With respect to RAM and RCS... we are still attempting to determine the impact of loss of RAM.[8]

As Col Allen's candid comments above imply, much research, development, testing, and evaluation was required following limited initial operational capability. The full-scale development phase never really stopped but continued after limited initial operational capability as "full-capability development" under a new fixed price incentive contract negotiated at $617.6 million, which was eventually completed at a final price $611.4 million.[128] Even after the completion of full-capability development, however, upgrade effort has continued. Another important continuing activity was RCS retention and repeatability testing to develop confidence that each production aircraft met the signature goals demonstrated in the initial test program and would continue to meet those goals throughout continued training and operations.

In the spring of 1984, shortly after limited initial operational capability, Col Howell Estes III became commander of the 4450th Tactical Group. He led the group through its first operational readiness inspection, which was successfully completed. According to Sherman Mullin, "The first ORI [operational readiness inspection] of the 4450th was a really major program milestone.... It went exceedingly well and the results were considered significant enough to be briefed to President Reagan. The most significant results were the large number of nearly perfect night bombing runs at U.S. Air Force ranges in the western United States."[181]

Post-Initial Operational Capability Testing and Development

One of the most important continued development activities after limited initial operational capability was associated with development of a precision guided munition, which complemented the unique nature of the F-117A and maximized its lethality. It was important to the capability of the F-117A as a complete weapon system and also represented the practical, problem-solving approach characterizing the Senior Trend program as a whole.

Precision Guided Munition Development

The precision guided munition (PGM) initially specified for the F-117A at initial operational capability was the GBU-10 laser guided bomb, which utilized a 2000 lb MK-84 body, a fin unit on the tail and a Paveway II guidance kit on the nose. Course corrections were made through the use of movable canards that were part of the nose-mounted guidance package. This weapon, introduced during the later stages of the Vietnam War to provide precision bombing against bridges and small targets, had two basic deficiencies related to its guidance system and the MK-84 body, which compromised the weapon's ability to precisely strike hardened targets.

The Paveway II guidance kit attempted to correct trajectory errors (angular errors between the velocity vector and the line-of-sight to the laser spot) through full deflection commands to the canards. These "bang-bang" control corrections led to overshoots about the optimum trajectory, which bled away energy, often resulting in a flat, slow terminal approach to the target, beneath the desired flight path. Under these conditions, the bomb, *if it even reached the target*, frequently had a high angle of attack, low impact velocity and poor impact angle— factors that degraded its accuracy and its ability to penetrate harder targets, as seen in Fig. 22.

Even with a good hit, the GBU-10 had limitations. The shape and integrity of the MK-84 general purpose bomb body was not suitable for penetrating buried, reinforced concrete targets. Ricochets, broaches, and premature breakup of the case were not uncommon. Furthermore, the forward and aft fuses could not withstand the massive impact deceleration to permit a time-delayed detonation *inside* the target. Considering the highly specialized niche that the F-117A was intended to fill by flying into heavily defended airspace to destroy high-value targets, these deficiencies would have seriously detracted from the F-117A's effectiveness.

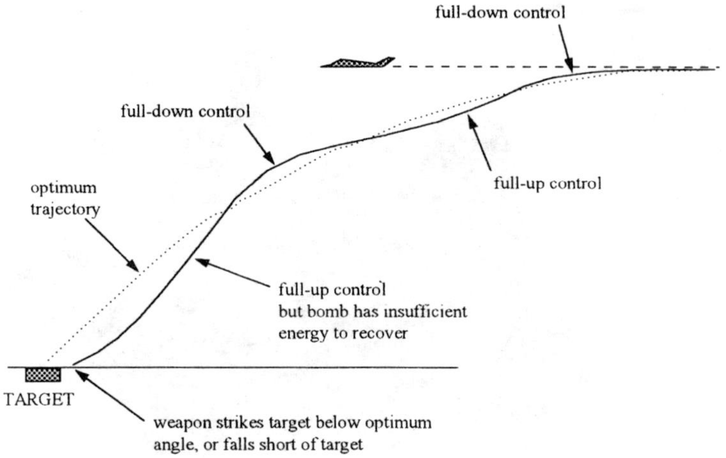

Fig. 22 Typical trajectory, "full-deflection" guidance.

Fortunately, ongoing initiatives enabled fielding of a weapon that better matched the F-117A's needs. First, the GBU-24 program, managed out of the U.S. Air Force Armament Division at Eglin Air Force Base, sought to correct the guidance deficiencies of the GBU-10 with the introduction of the new Paveway III guidance system by Texas Instruments, plus longer-span fins for improved glide efficiency. These kits used a proportional guidance algorithm, which maintained constant line-of-sight errors (constant angle between the velocity vector and the line-of-sight to the laser spot) through commands that deflected the canards only as much as required. This approach provided better energy management and improved accuracy when compared to the GBU-10, as shown in Fig. 23.

The guidance kit also permitted three different delivery modes: direct attack (a conventional trajectory primarily for use against vertical targets such as bridge towers), hard tuck (using a commanded pitch-down after launch to provide a steeper trajectory and improved impact angles against horizontal targets, such as the tops of bunkers), and loft (which offered a stand-off delivery, particularly with the improved glide performance offered by the GBU-24's larger tail fins), as seen in Fig. 24.[190]

The second PGM initiative involved a quick reaction capability program called Have Void, which was developing an improved 2000 lb bomb body with hardened target penetration capability, the BLU-109. This bomb featured a thicker case, integral aft section construction, and bolt-on bomb lugs (versus the traditional drilled holes for threaded lugs that compromised the structural integrity of the case). All of these features helped keep the bomb case intact until after the weapon had penetrated the target. An improved nose and body shape avoided ricochets and enhanced penetration. To provide

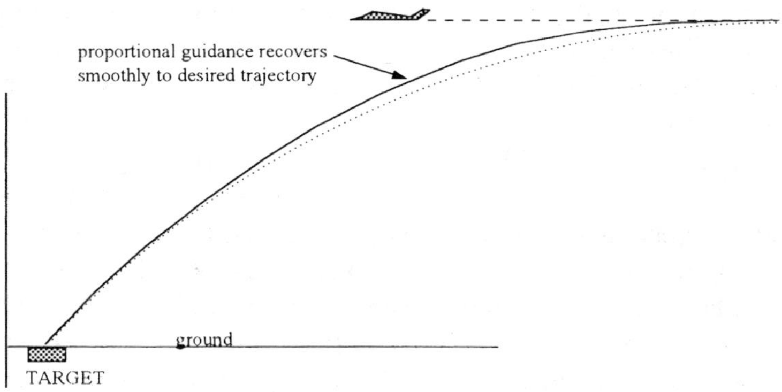

Fig. 23 Typical trajectory, proportional guidance.

consistent delayed detonation, the single fuse was on the aft end only and was of an improved high-*g*, high-reliability design. This combination of features ensured a penetration kill capability against a wider set of targets and over a broader range of delivery and impact conditions.[190]

The improved penetration capability of the BLU-109 coupled with the greater accuracy of the GBU-24 guidance package offered an excellent opportunity to improve the F-117A's lethality. Unfortunately, the GBU-24 would not fit in the internal weapons bays; the tail fins were too large and the adapter connecting the guidance section to the bomb body was too long.[49, 115, 190]

However, around 1985 three individuals involved in the F-117A weapons integration effort developed a concept that led to a new weapon providing the performance of a GBU-24 and that would fit in the F-117A's internal bays. This team consisted of Capt Charles (Chuck) Pinney, Combined Test Force (CTF) project manager for avionics and weapons integration; Ray Little, Lockheed ADP consultant for weapon separation and accuracy analyses; and MSgt Roland (Pete) Peterson, CTF crew chief for armament. Together, they performed an exploratory "fit check" to determine if the mating of the GBU-24's guidance section (with a shortened adapter), the MK-84 bomb body (the

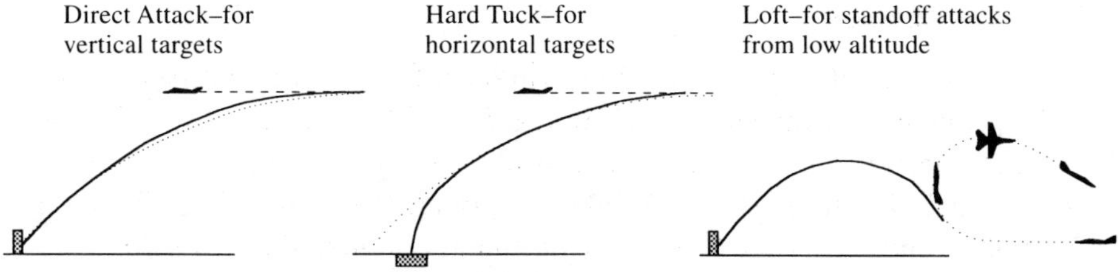

Fig. 24 GBU-24 delivery modes.

BLU-109 was not yet available), and the GBU-10's tail section could provide a "weapon" that would fit in the bay and clear all hardware during trapeze operations. The results of this "rapid prototyping" effort quickly won support and approval of the F-117A SPO and soon a new weapon, dubbed the "GBU-XX," was born.[190]

Immediately, the F-117A weapons team, a small group within the GBU-24 SPO (which was assigned responsibility for the proposed GBU-XX full-scale development and was directed by Col Ed Lippencot), and Texas Instruments began discussing how to modifying the GBU-24's seeker to accommodate the F-117A's mission and delivery profiles. Very few individuals were briefed beyond the GBU-XX effort into the Senior Trend program itself. The hard tuck and direct attack delivery modes complemented the F-117A's concept of employment, but the loft mode was deleted as it was inconsistent with the F-117A's straight and level delivery profile. Capt Pinney wrote the two-page operational requirements document, which Tactical Air Command endorsed, and the effort was underway. Early in this activity, the BLU-109 body, nearing completion in its development, replaced the MK-84 bomb body in comprising the GBU-XX weapon.[190]

As the effort became better defined, the weapon received the official designation GBU-27. Initial sled tests against concrete targets were very impressive. After successfully completing an F-4 flight test program, testing began on the F-117A by early 1987. The first accuracy test from an F-117A gave an early indication of the GBU-27's precision: the inert weapon scored a direct hit on the 55 gallon barrel being used as a target (which was heated with charcoal to simplify the acquisition and targeting tasks), splitting the barrel in half.[190] Certification testing of the GBU-27 with the F-117A was completed in late 1988.

The F-117A/GBU-27 combination made its operational debut in Desert Storm in 1991, demonstrating unprecedented accuracy (achieving over 95% probability of a hit, according to the placard on the GBU-27 exhibit at the U.S. Air Force Museum near Dayton, Ohio) and ushering in a new era in precision strike capability. The F-117A's ability to reach heavily defended, high-value targets, coupled with the GBU-27's accuracy and penetration capability, created a weapon system with tremendous military value.[77]

Air-to-Air Missiles

Air-to-air capability (AIM-9 Sidewinder infrared guided missile) was initially proposed, primarily to allow the F-117A to be used against Soviet AWACS aircraft.[37, 78, 142] Air-to-air missile testing was therefore included in the original (February 1980) Flight Test Plan.[179] In July 1982, Tactical Air Command and Air Force Systems Command conducted a quick-look assessment of the technical issues associated with target detection, recognition, and

tracking on a hypothetical mission that was "not well defined, but...designed to stimulate some creative thought."[37] The proposed operational concept was tenuous, involving long-range missions with multiple air refuelings. The response from the F-117A program office, delivered within two days, concluded that even with the help of friendly AWACS, there would be significant issues with locating, identifying, and tracking the target while remaining completely passive. Soviet fighter threats to the U.S. AWACS and tankers, specifically MiG-31s, would have possibly necessitated CAP protection, which in turn would require additional tankers. "Thus a minor task force must be assembled for each of these 10- to 11-h missions."[231] The SPO noted that the F-117A would definitely not have air-to-air capability at initial operational capability because the necessary testing would not yet have been accomplished.

Air-to-air capability was not part of the definition of limited initial operational capability to which all pre-initial operational capability testing was oriented; so any missile testing would only be conducted after initial operational capability. In fact, such testing was never conducted. The only air-to-air missile testing actually performed was static form and fit testing of the AIM-9 on the F-117A.[190] Some testing with the AIM-7 Sparrow semi-active radar-guided missile had also been planned. However, to employ the AIM-7, the F-117A would have required a complex radar system. Although Lockheed had developed a new low observable radar antenna concept, this was not pursued,* and so the F-117A was incapable of using the AIM-7 missile in any case. Neither missile was ever launched from an F-117A in flight.[190]

*A concept involving the use of the B-1 radar (built by Westinghouse) to provide the F-117A with synthetic aperture radar for use in all-weather missions was installed and flight tested on aircraft number 784. The radar was tested with a unique "evanescent radome," patented by Lockheed's David Young, which was designed to maintain the F-117A's low radar signature. The design proved too costly and was not adopted on the F-117A; however it was the direct predecessor of the radome and integrated forebody now on the F-22.[181, 212]

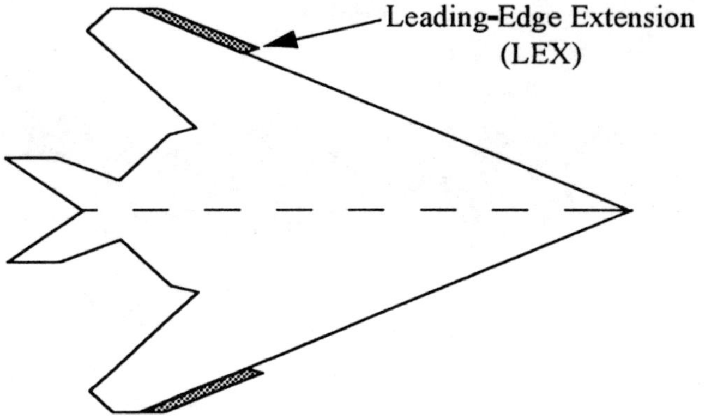

Fig. 25 Leading-edge extension, tested late 1983/early 1984.[175]

Leading-Edge Extension (LEX)

A wing leading-edge extension (illustrated in Fig. 25) was added to Aircraft 780 while it was on the ground from September through November 1983 and tested from December 1983 to February 1984. This was intended to improve low-speed handling qualities and reduce the landing approach speed. The LEX was not adopted; the aerodynamic benefits were found to be too small to warrant a fleet retrofit.[175, 190]

Navy Evaluation

A "quick-look" evaluation in late 1984 explored issues associated with carrier operations, primarily low-speed handling qualities.[175, 232] Two Navy pilots each made three to six flights in an F-117A. General conclusions were that the F-117A was not suitable for carrier operations.[190] In addition to having a very high landing approach speed, which was reduced only incrementally by the LEX, the F-117A structure was not suitable for catapult takeoffs and arrested landings. The aircraft would have needed major structural changes to include a strengthened fuselage, new landing gear, and a new tailhook. Also, there were not at that time viable solutions to the problems of supporting and maintaining a low observable aircraft in a maritime environment. Repair facilities on board ship were not equipped to handle the F-117A's special edges and coatings. Environmental factors including salt spray and "dings," which are hard to avoid on crowded carrier decks, could adversely affect the special materials and critical smoothness required for low signature. Ten years later, solutions to these problems appear to be coming within reach. However, this does not change the fact that the F-117A design is still basically unsuitable for carrier operations.

Tail Failure

Aircraft 781 lost one of its tail fins on September 25, 1985.[232] Number 781 was allocated to performance and flying qualities evaluations, as well as all-weather and weapons separation testing (previously briefly used for RCS testing). The aircraft had just completed the exploration of all planned flutter/loads data points and was beginning a "piggy back" weapons compatibility test, when the left fin failed following a pull-up maneuver at 10,000 ft altitude. The incident was filmed by a chase aircraft. The F-117A pilot, Maj John Beesley, successfully landed the aircraft and was covertly awarded the Distinguished Flying Cross by a senior Air Force general officer.

In the opinion of Lt Col Stan Siefke, U.S. Air Force (who as a captain was assigned to the F-117A CTF and was responsible for aircraft performance and weapons separation testing),

> It was the many hours of dedicated test planning and simulator runs that were accomplished during the high angle of attack [AOA] test program that ensured that the aircraft was not lost during the incident. Many hours of simulation runs had been conducted by Maj Beesley and the Lockheed test pilot, Dave Ferguson, under the guidance of Bob Loschke (Lockheed flight controls engineer), Maj Page McGirr (Air Force Flight Test Center flight test engineer for high AOA and aircraft handling qualities), and myself during the development of the F-117A flight control computer (FLCC) and the clearing of the high AOA envelope. Every conceivable failure mode was tested in the [Lockheed Rye Canyon] simulator, including the loss of elevons, vertical fins and multiple channels of the FLCC to ensure that everyone knew what to do in the event of a failure. Every flight test data point in the high AOA program was used to continuously update the Rye Canyon simulation following each mission. All of this caution was required since we were accomplishing the high AOA program without the use of a departure recovery parachute (which is normally the case in every high AOA test program conducted at Edwards Air Force Base. However, due to the unique nature of the design of the F-117A it was felt that a recovery chute would unduly influence the aerodynamic effects of the vehicle we were trying to measure and thus wrongly influence the design of the FLCC.)
>
> I was the test conductor in the mission control room the day A/C 781 lost its left vertical fin, and everything performed as the simulator predicted it would. Everyone knew what to do and when including slowing

> the aircraft down, gaining altitude to get out of the high dynamic pressure condition and lowering the landing gear to increase the directional authority of the FLCC for the remaining fin. Later John [Beesley] commented that he felt the vibration as the fin left the aircraft, heard Dave Ferguson's radio call from the chase aircraft that the event had occurred, but never knew the difference in the flying qualities of the aircraft during the uneventful landing.[212]

As noted, Maj Beesley had previously "flown" the F-117A in the "single tail" configuration in the Lockheed simulator at their Rye Canyon, California, research and development facility as part of an exploration of potential flight control system failure modes.[190] Piloted simulations are an important tool to prepare pilots for structural and systems failures, as well as for prediction errors (in other words "what if the aircraft's directional stability is less than we think it will be?"). Several potential mishaps were avoided in this way during the Have Blue and Senior Trend programs.

This event revealed a previously hidden problem. Analysis following the incident showed that a critical flutter mode was only damped by rudder post bearing friction. Although the friction had suppressed the flutter mode during all prior testing, the friction was highly dependent on the fin loads. Loads on the fin during the pull-up probably deformed the fin/rudder structure in a way that relieved the bearing friction, causing the particular flutter mode to become divergent. This discovery forced the issue of the long-awaited fin redesign, which was authorized in August 1986. The new fin, made of graphite-thermoplastic composites, was first flown on Aircraft 784 (the fifth full-scale development aircraft) in July 1989. Some of the later production F-117A's incorporated the thermoplastic fin during production. The remainder were retrofitted, and by 1992 all F-117As had the new composite fins.[175, 190, 232]

Offensive Capability Improvement Program (OCIP)

The first phase of the OCIP consisted of a weapon system computational subsystem upgrade begun in 1984. The F-117A initially relied on three Delco M362F computers for weapons delivery, flight control, navigation, and other functions. The upgrade program replaced the Delco computers with IBM AP-102s, providing a large increase in processing capability.[104, 153] All F-117As had received the upgrade by 1992.[49, 104]

Phase II of the OCIP used the added computational capability to provide improved situational awareness and reduced workload for the pilot. Improvements included a four-dimensional flight management system (three spatial coordinates, plus time on waypoint to within ± 1 s), revised cockpit with Honeywell multifunction displays, and a digital tactical situation display or "moving map," a new data entry panel, improved autopilot, auto throttle, and pilot-activated attitude recovery system. The first flight of a modified F-117A was made on December 1, 1988, with the first service redelivery in November 1990. Redeliveries proceeded at a rate of approximately one per month,[49, 104] All F-117As in service presumably have received the OCIP phase II upgrades.

Phase III included a new IRADS by Texas Instruments, offering increased detection range as well as longer life for the laser. A Honeywell ring laser gyro inertial navigation system replaced the original SPN-GEANS INS (also by Honeywell), which had been adapted from the B-52 but is no longer in production and is expensive to maintain (although its performance and reliability were better than initially expected).[31] The INS will be supplemented by a global positioning system unit.[104]

Overall Follow-on Development Costs

Continued F-117A development and upgrades from 1984 to 1990 reportedly cost $1.3 billion (in FY93 dollars), which is roughly equal to the amount invested for the full-scale development program from 1978 to 1984. Total development cost up to 1990 was $2.5 billion (in FY93 dollars). Approximately half of that was initial development (full-scale development), one third follow-on development, and one sixth upgrades.[179] This represents an unusually large follow-on effort and is primarily the result of the following causes:

1) The initial strategy was to create an operational system as quickly as possible.

2) Limited initial operational capability requirements were specifically defined in advance with full capability to be achieved subsequent to initial operational capability.

3) Unforeseen technical problems added to the required follow-on effort.

4) The increased size of the procurement required more emphasis to be placed on reliability and maintainability than originally planned.

All aircraft experience some post-EMD improvements and modifications to varying degrees, driven by shortcomings in the original designs, by changes in technology, and by changes in requirements. Since it is almost impossible to forecast any of these events, the approach of fielding a limited capability quickly and following up with modifications and upgrades may be the most cost-effective in some cases.

Reliability and Maintainability

Reliability and maintainability of F-117A was initially very poor compared to other tactical aircraft. Various special low observable features—surface coatings, aperture treatments, special air data probes, and so on—had not been used before on operational aircraft, and were initially very maintenance-intensive. Also, overall design requirements for low observables constrained the layout of the aircraft and its subsystems. The best arrangement for stealth was not necessarily the best arrangement for maintenance accessibility. Early program strategy was to accept these penalties rather than trying to solve all of the low observable reliability and maintainability issues up front. This was not an oversight but a conscious decision to keep the program focused, to field the system as quickly as possible, and above all to minimize the risk of not achieving very low observables capability at all.

Experiences with Have Blue had indicated that doors could dramatically increase the vehicle's RCS if not properly treated or completely closed. When this happened in the Have Blue program, the test team could bring back the airplane, inspect, fix the problem, and repeat the test. This would not be the case if a strike aircraft were sent into combat with its RCS several times higher than it was supposed to be. For this reason, the Air Force wanted to be extremely conservative with doors and access panels on the advanced technology aircraft (ATA—the early designation for the F-117A). An early goal was to have only four openings on the aircraft: the three landing gear doors, plus the cockpit.[34] All maintenance would be performed through one of these four openings. Obviously, as detail design progressed, it became apparent that some additional access had to be provided. The overall approach, however, was that if the aircraft wasn't stealthy, it did not matter how maintainable it was because it would not be able to perform its mission. Airframe openings were seen as a risk to stealth, and so they were strictly minimized.[75, 34]

As a result, the F-117A's maintenance man-hours per flight hour were extremely high at limited initial operational capability with monthly averages typically around 150 maintenance man-hours per flight hour. For comparison, maintenance man-hours per flight hour at that time were approximately 32.3 for the F-15A/B, 22.1 for the F-15C/D, and 19.2 for the F-16A.[64, 127] Operating costs of the F-117A were estimated to be roughly double the dollars per flight hour of the F-15 or F-4G and three times as much as the F-16A.[215]

Reliability and maintainability performance for a "good" month shortly after limited initial operational capability is summarized in Table 21. The mission capable rate represents the percentage of aircraft returning from a sortie with no malfunctions of mission-critical systems. Full mission capable rates refer to the percentage of aircraft returning from a sortie with no malfunctions that would limit mission performance in any way. In the early period of F-117A operation, the condition of the low observable aspects of the aircraft was not factored into these rates but was accounted for in a separate category created specifically for the F-117A. Full mission capable, stealth represents the percentage of aircraft returning from a sortie with no known defects or damage to the low observable coatings, access panel treatments, and so on.[8] The full mission capable stealth rate was initially very low, as shown in Table 21.

The figures fluctuated considerably from month to month, because of the small number of units and the still-maturing state of the system. Maintenance man-hours per flight hour had actually been 164 during the first complete month of operation (November 1983); January 1984 was a "good" month at 112.7, and the average for February was 127.[127] Even these monthly averages showed considerable "smoothing-out" of day-to-day variations. Maintenance man-hours per flight hour at times was over 200, according to Brig Gen James Allen, U.S. Air Force (Ret).[7]

Although the 4450th Tactical Group was closely involved with the evolution of the F-117A and its maintenance procedures, Tactical Air Command Headquarters' expectations were not initially consistent with the realities of the system that was being developed. The 1983 version of the *System Operational Concept for the ATA* had stated a goal of 37 direct maintenance man-hours per flight hour at the organizational level, with the expectation that this would drop to 22 maintenance man-hours per flight hour "for a totally mature system" (three years after initial operational capability).[78] The 1985 *System Operational Concept* was more realistic, stating a goal of 77 maintenance man-hours per flight hour at maturity.[79] The 1987 *System Operational Concept* did not state a specific reliability and maintainability goal, but noted that "the minimization of the maintenance burden, expressed in maintenance man-hours per flight hour, will be a high priority goal of the F-117 contractor."[80]

The initial approach to reliability and maintainability was consistent with the original concept of a small number of aircraft (around 20) supported by a contractor-operated depot level maintenance facility at Lockheed Burbank.[240] However, this approach became less cost-effective as the F-117A procurement quantity was increased. Changes were made in the maintenance concept including increased intermediate level maintenance with the Air Force assuming a greater share of the maintenance burden. To do this, the Air Force

Table 21 Early reliability and maintainability performance of the F-117A[127]

	Best month to date (Jan. 1984)
Sorties per month per A/C	7.5
Flight hours per month per A/C	11.2
Mission capable (MC) rate	59.5%
Full mission capable (FMC) rate	31.8%
Full mission capable, stealth (FMC/S) rate	10.8%
Maintenance man-hours per flight hours	112.7

required technical data that had not been included in the original contracts with Lockheed. Therefore, considerable data had to be procured later and at greater cost than if they had been procured up front.[47, 53]

At the same time, as the production run was extended, the need for reliability and maintainability improvements became even more acutely apparent as flight operations began. As Joseph Reiman, a former F-117A SPO engineer recently recalled, "At one point Gen Creech [commander of Tactical Air Command] said that he wouldn't support funding any aircraft past the original 20 unless we did something to improve reliability and maintainability. This resulted in a MIP (maintainability improvement program) and a RIP (reliability improvement program) with specific improvement goals."[194] Modifications were made to the aircraft in an effort to improve its reliability and maintainability. The contractor had, in fact, been working to address various reliability and maintainability issues since very early in the program. However, in many cases workable, robust solutions took a long time to achieve. One example is the spray-on RAM coating process (which affects both the producibility and the maintainability of the F-117A). Lockheed began to experiment with RAM spraying in 1980, but it was 1985 before this could be used routinely on major portions of production airframes.[34, 232] By 1989, maintenance had been reduced to about 45 maintenance man-hours per flight hour.[64, 215] However, this improvement and comparable improvements in mission capable rates and mission reliability, for example, were only obtained through a Herculean effort.

Furthermore, by 1989 the direct maintenance specifically charged to low observables had dropped to less than 10% of the total, indicating that the problems of maintaining the coatings had been addressed fairly effectively.[64, 215] However, the low observable work unit code only includes specific items, and does not completely reflect the reliability and maintainability impact of many design characteristics that were necessary to make the F-117A a very low observable weapon system. In spite of continued effort and expenditure, there has been little improvement in the F-117A's reliability and maintainability statistics since 1989.[215] The implication is that once a design is set, certain intrinsic characteristics cannot be readily changed. The original program philosophy of *stealth at all costs* was necessary to keep the effort focused and maximize the probability of successfully introducing a dramatic new capability into the operational inventory. However, the F-117A will probably always be more maintenance-intensive than its nonstealthy counterparts.

Brief Operational History

The original operational F-117A unit, the 4450th Tactical Group, was renamed the 37th Tactical Fighter Wing (TFW) of the 12th Air Force on October 5, 1989. The Tactical prefix was deleted for all Air Force fighter units on November 1,1991, as a consequence of the integration of the former Tactical Air Command and Strategic Air Command into the new Air Combat Command. Starting in 1992, consistent with the relaxation of some of the security aspects of the program, F-117A operations were moved from Tonopah to Holloman Air Force Base, New Mexico. Aircraft 791 was delivered to Holloman on Janurary 7, 1992, for maintenance familiarization, and movement of the remaining aircraft took place between May 9 and July 5, 1992, at which time the unit became the 49th Fighter Wing.[49, 104] F-117A operational unit commanders are listed in Table 22.

Table 22 F-117A operational commanders through 1993[49]

Name	Unit Designation	From	To
Col Robert A. Jackson	4450th Tactical Group	Oct. 15, 1979	May 17, 1982
Col James S. Allen	4450th Tactical Group	May 17, 1982	June 15, 1984
Col Howell M. Estes	4450th Tactical Group	June 15, 1984	Dec. 6, 1985
Col Michael W. Harris	4450th Tactical Group	Dec. 6, 1985	April 3, 1987
Col Michael Short	4450th Tactical Group	April 3, 1987	Aug. 10, 1988
Col Anthony J. Tolin	4450th Tactical Group	Aug. 10, 1988	Oct. 5, 1989
Col Anthony J. Tolin	37th Tactical Fighter Wing	Oct. 5, 1989	Aug. 16, 1990
Col Alton C. Whitley, Jr.	37th Tactical Fighter Wing	Aug. 16, 1990	Oct. 1, 1991
Col Alton C. Whitley, Jr.	37th Fighter Wing	Oct. 1, 1991	July 5, 1992
Col Alton C. Whitley, Jr.	49th Fighter Wing	July 5, 1992	July 8, 1992
Brig Gen Lloyd W. Newton	49th Fighter Wing	July 8, 1992	June 1993

The first combat use of F-117As occurred while the F-117As, under the command of Col Tony Tolin, were still at Tonopah. This occurred on December 21, 1989, during Operation Just Cause in Panama. Two F-117A's each dropped a 2000 lb bomb in the barracks area of the Panama Defense Force at Rio Hato as a diversionary tactic. "This operation, though successful, did not test the aircraft from a low-observables standpoint as Panama had no radar defense network."[175]

The second, much more significant use of the F-117A in combat took place during Operation Desert Storm against Iraq from January 17 to February 28, 1991, under the command of Col Al Whitley, although Col Tolin was still the wing commander when the deployment to Saudi Arabia began. More than 40 F-117As flew a combined total of 1270 combat sorties. They dropped 2087 bombs, achieving 1669 direct hits.[49, 77] F-117As were the only manned aircraft to attack heavily defended targets in Baghdad. Not a single F-117A was hit by enemy fire.[175]

The F-117A achieved a mission capable rate of more than 85% during Desert Storm. An Air Force briefing titled "The Gulf War—A Survey" noted that the F-117A was the only aircraft capable of hitting hardened targets in extremely high-threat areas.[77] Although F-117As flew only 2% of the strike sorties expended during all of Operation Desert Storm, their unique capabilities resulted in their use against 40% of all the Iraqi strategic targets.[77]

The number of sorties flown by the F-117As during this operation corresponds to a rate of slightly less than one sortie per night per aircraft. An intensive maintenance and logistical support effort was undoubtedly required to sustain that rate. However, it might be useful to think in terms of maintenance man-hours per target hit, rather than maintenance man-hours per flight

hour. The F-117A generally requires only tanker support; conventional strike aircraft may require fighter escort as well as lethal (Wild Weasel) and nonlethal (jamming) suppression of enemy air defenses, plus additional tankers for all of the above to strike the same target. The cumulative maintenance necessary for such a force would make the F-117A's maintenance statistics look good. However, the F-117A is not invulnerable, and it may at times need support by employing jamming, decoys, or other types of support to enhance its effectiveness.

The experience of Desert Storm bears out the assertion, voiced by proponents of the increased procurement during the early part of the 1980s, that the ATA (as it was known at the time) would be a valuable asset in a major conflict.[78, 130, 145, 240–243]

Chapter 7

Senior Trend: Lessons Learned

Requirements (primarily from Chapters 3 and 4)

Narrow focus—achieve a very low observable system quickly and secretly—was the key to program success. All other performance goals were traded off when necessary to preserve the primary program objectives.

Requirements developed for the F-117A were based on the successful demonstration of reduced signatures on Have Blue and were primarily focused on achieving a similar signature reduction on an operational military aircraft rather than trying to push ahead to "next generation" low observables technology. The "straight and level" employment concept adopted for the aircraft on its primary night attack mission profile allowed all performance goals except signatures to be quite moderate.

To this end, requirements were kept simple and realistic. "For example, all weather attack capability using a high resolution radar with a crew of two was [initially] desired by the users. Requirement tradeoffs were made that considered the impacts of the active radar sensor on the low observable characteristics of the

vehicle...."[87] Night, clear air capability was attainable, and this became the requirement.

Air-to-air combat capability was also initially desired by some members of the user community. As late as May 1983, the system operational concept for the ATA stated that, "As a collateral task, the F-117 can be used as a clear air interceptor against selected, high-value targets such as [Soviet Union] AWACS and airborne jammers."[78] This reflects the gap that can exist between the expectations of the user and what is technically feasible. The F-117A program successfully focused on achieving the essential basic capability and resisted pressure to add multimission capabilities that were not supported by the available technology. This focus was a key element to the program's success. Later versions of the system operational concept were progressively more realistic in their assessment of the aircraft's capabilities in several areas.[79, 80] Sometimes it pays to know when not to add capabilities, to keep the system development focused and affordable.

Also, the conservatism in developing a relatively small attack aircraft such as ATA "A," rather than a larger bomber ATA "B," proved to be a wise choice. Development difficulties with the F-117A, particularly with the exhaust system and the airframe structure, suggest that the "B" would indeed have been too risky because of its significantly higher gross weight and partially afterburning engines.

As development progressed, flexibility was retained to "make modest tradeoffs in capability for significant payback," as long as these tradeoffs were in areas that did not have a serious impact on the aircraft's ability to perform its mission. For example, the initial load factor of 7.33 *g* was relaxed to 6 *g* to avoid structural weight growth.[87] The limit load factor was reduced to 4 *g* at initial operational capability, pending a wing structure redesign.[179] This is another example of flexibility; the limitation was accepted for a time, rather than delaying the whole program. Nevertheless, such flexibility should only be exercised with noncritical requirements.

DEVELOPMENT AND TESTING (CHAPTER 4)

Advances in radar cross section (RCS) prediction techniques helped designers to *understand* how to design low observable aircraft, contributing to a breakthrough. The successful realization of this breakthrough, however, still relied heavily on experimental measurement. As signatures were reduced to unprecedented levels, serious shortfalls were identified in the ability of U.S. industry and government test ranges to measure very low signatures.[186] An intensive effort to address this issue progressed in parallel with the Have Blue and Senior Trend programs. Subsequent low observable weapon system development programs have benefited tremendously from the progress that was made in this area.

Modeling and prediction techniques—whether in electromagnetics, aerodynamics, structures and materials, or other disciplines—are governed by computational power and an understanding and mathematical description of the underlying physics. Recent advances in computational power have created a tendency to believe that computational modeling can replace experimental measurement. However, advanced weapon system designs are almost always near the limit of some aspect of our understanding of physical phenomena. Therefore, experimental techniques will always be an important complement to computational methods in scientific research and in the design and development of advanced weapon systems.

Alan Brown, who became the Lockheed program manager for the F-117A (then referred to as the ATA) at the time of full-scale development start, had little prior involvement in the ATA design studies, having been fully committed to Have Blue instead. However, Jack Twigg had strongly urged that Brown, the primary low observables expert, be appointed as Lockheed's ATA program manager for the full-scale development phase rather than someone more experienced in traditional aircraft design and management.[34] Programs that involve new technology need someone in charge who understands and appreciates the new technology, particularly when the technology is risky or controversial.

While developing the F-117A, the decision not to push too many technologies at once was critical. Effective use was made of off-the-shelf subsystems and components in the F-117A development program to reduce cost and technical risk and to expedite development and fielding of an operational system. Col Ted Bowlds recently noted:

> The F-117A was primarily a collection of off-the-shelf components. This was especially true of the avionics. The big advantage of developing a weapon system from this approach is the availability of fully tested and in production subsystems. The F-117A did not have to design, test, and integrate subsystems like the FLIR. Rather, the F-117A only had to integrate these subsystems. Depending on your experience however, this equated to half of the problem already being solved. The integration proved to be a challenge. Most of the avionics subsystems used a different coordinate reference system. Some used latitude, longitude, altitude while others used north, east, down or north, east, up. Additionally, the protocols were not necessarily the same. As a result, one of the weapons systems computer's main tasks was to serve as avionics subsystem integrator. It had to convert the various coordinate systems into a common one before computation. This added burden consumed some extra effort, but as most problems on the F-117A, it was overcome.[31]

This approach, despite the integration difficulties, reduced development time, cost and risk. Additionally, production savings were realized because this gave subcontractors and vendors economy of scale.

Weapons integration tended to lag aircraft development. Weapons integration activities continued for several more years after limited initial operational capability. In addition, the weapon that seemed ideally suited to the F-117A's concept of operations (GBU-24), was under development concurrently with the F-117A, but *would not fit in the F-117A's internal weapons bays*. An aggressive effort was undertaken by a small number of individuals to address this situation. This led to the GBU-27 project. The nature of the GBU-27 concept definition and development mirrored much of the Skunk Works philosophy and success shown in other areas of the Have Blue and Senior Trend programs:

1) Program managers recognized current deficiencies and the opportunities to leverage ongoing efforts (GBU-24 and BLU-109) to correct those deficiencies.

2) Highly motivated people closest to the problem were empowered to work key issues, providing excellent results in minimal time.

3) Streamlining the acquisition process and limiting bureaucracy helped to achieve capability in the shortest possible time.

4) Unhindered by bureaucratic barriers, informal networks between the various government agencies and the contractors facilitated quick response and a high degree of teamwork at the working level.

5) Established Air Force practice prohibited developing a weapon specifically for a single aircraft. As it turned out, the GBU-27 effort not only provided the F-117A with vastly improved lethality, but it has since been employed on other aircraft as well, including the F-15E Strike Eagle. The smaller fins, relative to the GBU-24, allow for more compact carriage, which has

been found to be a useful feature for missions that do not specifically require the superior glide performance of the GBU-24.

Fly-by-wire technology proved to be very transferable. The relaxed-stability design adopted for the F-117A depended on active fly-by-wire control. The F-117A flight control system was based on the F-16's fly-by-wire system. However, fly-by-wire was not and is not a panacea for every stability and control problem. It cannot always compensate for faulty vehicle design. Directional stability problems, encountered during the testing and development of the F-117A, could not be solved by any amount of automatic control compensation but rather required changes in aircraft design.[9, 10, 54–61] Attention must be paid to the handling qualities that are desired, and the aircraft must be designed in parallel with its control system to ensure that those handling qualities are obtained.

Control system development must proceed with analysis, simulation, and careful flight testing. As development progresses, the effects of any flight control parameter change must be thoroughly evaluated to ensure that there will be no adverse side effects. Analytical and simulation capabilities should be available during the flight test program (on-site if possible, although this was not the case for Have Blue and Senior Trend, which relied heavily on simulators at Lockheed's Rye Canyon research facility). Simulations should be conducted to prepare pilots not only for the expected flying qualities of the aircraft but also for unexpected flying qualities, as well as structural and systems failures. On more than one occasion, including the directional instability on the first flight and the fin failure during later weapons compatibility testing, pilots were able to take the correct actions during critical situations because they had flown piloted simulations of the conditions encountered.

Throughout development and testing, the teamwork between the contractor, the Air Force Systems Command personnel, and the Tactical Air Command personnel was crucial to achieving success, particularly in view of the rapid pace of the program. Operational people—pilots and maintainers—were necessarily involved very early and worked closely with the contractor and the SPO. The operational unit was formed at almost exactly the same time as the initial long-lead production contract was signed (November or December 1979), —only one year into the full-scale development program. According to Brig Gen James Allen, U.S. Air Force (Ret), who as a colonel commanded the operational F-117A unit at the time of initial operational capability,

> The operational maintenance personnel were deeply involved from the early days. They formulated the sortie generation concepts, shared their personal experience of off the shelf parts with the contractor, and ensured the effectiveness and deployability of maintenance vans. In technical publication the operational unit made the decision to eliminate illustrated parts breakdown manuals, use Navy tech data for the GE 404

> engine, and actually wrote many of the job guides. The joint validation/verification teams reduced the process time in half....[7]

The highly concurrent, collocated test environment created some friction but led to an environment where effort could be focused on problem solving with a minimum of organizational or bureaucratic barriers. Col Ted Bowlds noted

> From my perspective, the reason that the F-117A acquisition worked is because the number of individuals involved was kept at the minimum. Very often, individuals were asked, most of the time they volunteered, to contribute their time and talent in an area outside their normal work responsibility. It was not uncommon to see someone other than the "regular crew chief" working on a project. Aerial photography is an excellent example of this approach.
>
> If you go to Edwards Air Force Base, the only people allowed to take pictures, for that matter take cameras, in an aircraft (the photo safety chase) are the designated photographers. This was not the case with F-117A testing. Flight test engineers, individuals who understood the technical details of a particular test, also took the necessary aerial pictures. This allowed for one individual to be responsible for both the technical interpretation of the test and also the visual aspects. The net result was fewer people.[31]

Allen further commented that

> All of us that were directly involved agreed and underscore that a key and essential element of the success experienced was the close daily interrelationships. As in any program there were conflicts and interpersonal disconnects but those were overcome.... I must point out that the collocation of training and testing was fundamental to the success in that the constant "cross ruff" between all parties in ops, maintenance, and support was absolutely necessary in the highly concurrent environment. It was not uncommon for pilots in training for the ops unit to reach test points in the flight envelope before the assigned test pilots did so.[7]

There were some difficulties and disagreements between the various organizations involved in the test program, as described earlier (Chapter 4), but most important, at the working level the contractors, testers, operators, maintainers and supporters were able to work together to overcome the many challenges to develop this unique weapon system.

Production (Chapter 5)

Although the production rate was very low (eight per year) the production cost of the F-117A was reasonable ($42.6 million unit flyaway cost).

Building in small numbers on a big assembly line under-utilizes facilities and is expensive. But a low production rate can work well if the facilities are scaled to the *realistic* maximum production rate, as was the case with the F-117A. The production facility was designed for a rate of eight to 12 aircraft per year, and the actual rate was in fact eight aircraft per year for most of the production run. According to Paul Kaminski, under secretary of defense for acquisition and technology, who was in charge of the Pentagon low observables project office during much of the early production phase, Senior Trend was "one of the few programs I have worked on where we built at close to the rate we had facilitized for." [115]

The F-117A experience showed that small production runs can achieve "learning curve" benefits comparable to larger production runs. The 20th production F-117A only required 42% as many man-hours to build as the first production aircraft.[99] On any size program, a learning curve will not be achieved if the production rate fluctuates, causing experienced workers to be lost when the rate drops and then be replaced by inexperienced workers when the rate rises again.

In the F-117A program, careful attention was paid to maintaining a relatively efficient rate and avoiding dips or, worse yet, gaps, in production. This was particularly critical during mid-to-late 1981, when the increased procurement of a full wing of F-117A aircraft was being considered. Long-lead funding was allocated to maintain an efficient rate and transition smoothly into the possible follow-on procurement. Excellent and highly focused cooperation between the contractors, the Air Force, Office of the Secretary of Defense, key Congressional staff members, and committee chairmen enabled rapid replanning decisions and allowed for needed reprogramming of funds to develop and implement the best possible delivery schedule for production efficiency and early initial operational capability.

The lesson is that timely authorization and funding of long-lead procurement is necessary to maintain production efficiency and to protect schedule. Commitment of long-lead money early reduces the probability of high production costs and timely schedules.

However, the F-117A production methods were very manpower intensive. Each aircraft was practically handmade, particularly in the early stages of the production run. This resulted in quality control problems and variability in the product, both of which contributed to the F-117A's initial very high maintenance man-hours per flight hour. The producibility improvement program focused on improving quality and reducing aircraft production man-hours. Some of the improvements, such as spray-on RAM versus the older glue-on sheets of RAM, reduced both production and maintenance man-hours. Other efforts focused on specific reliability and maintainability issues.

OPERATION AND SUPPORT (CHAPTER 6)

Lowering the F-117A's maintenance requirements to reasonable levels was a long and difficult process. Incentives created by the initial maintenance approach added to the difficulty. Because of the specialized nature of the system, and the fact that relatively few tech orders (maintenance and operation manuals) had been initially contracted, the Air Force was dependent on Lockheed for many of the specialized low observables maintenance tasks.[47, 53] This situation did not create a strong incentive for the contractor to reduce the aircraft's maintenance needs, thus, some continuing costs were incurred because of the initial decisions to rely on contractor maintenance. Paul Kaminski recently noted, "We [the Air Force] probably should have better incentivized the contractor in the supportability area."[116]

The lesson is that reducing costs in all sectors of national defense—research and development, procurement, operations and support, and others—will require appropriate incentives.[116]

Reliability and maintainability may have received earlier attention if initial program priorities had been different. However, this would have entailed additional design effort, development time, and expenditure. Designing for maintainability would have imposed additional constraints on the overall weapon system configuration and might have made it impossible, given the state of the art at that time, to achieve a viable very low observable aircraft design solution. Furthermore, in turning the design into a real aircraft, new technology and unique approaches were involved. The major challenge was to achieve solutions that worked at all. Testing and refinement of these solutions to achieve high reliability would have taken time, which would have been contrary to the initial goal of introducing the F-117A into the operational inventory as rapidly as possible.

Most of the F-117A's reliability and maintainability issues have been worked, and considerable improvements were achieved in the first five to seven years of service, many, however, at higher cost than would have been the case if they had been an early program priority. However, the F-117A is still more maintenance-intensive than comparable nonstealthy tactical aircraft.[215]

The lesson is that some characteristics will be intrinsic to the basic design of a weapon system and subsequently cannot be changed. As is widely appreciated today, up-front attention to reliability and maintainability is necessary to ensure a highly reliable, maintainable, and supportable weapon system. Future experience will answer the question of how far the reliability and maintainability penalties associated with low observable aircraft can be minimized through careful attention to this issue in the early stages of design and development.

However, it is important to remember that if *too many* considerations, sometimes referred to as "-ilities," are taken into account up front (such as producibility, reliability and maintainability, interoperability, environmental compatibility, competition, economic impact, and others) a program can lose focus and even become almost inexecutable. A high degree of focus was maintained in the Senior Trend program, a very low observable system was successfully developed, and an early initial operational capability was achieved, consistent with the original program priorities. The resulting weapon system has proven to be a valuable national asset in spite of its limitations.

Concurrency (Chapters 4 through 6)

Extent of Concurrency in the Senior Trend Program

The F-117 program was highly concurrent by modern standards. There were less than four years between program go-ahead (November 1978) and planned initial operational capability (March 1982). Limited initial operational capability was actually achieved in October 1983, later than planned but still only five years from program go-ahead. The first production lot contract (for seven aircraft) was signed several months prior to the first flight of the full-scale development program. The first five production lots—28 aircraft, or approximately half the production run—were on contract prior to initial operational capability.[232] "Concurrency can be expensive if problems emerge during test flights and result in major modifications to dozens of complete and near-complete aircraft." [229] or, worse yet, the aircraft proves unable to perform its mission after significant production funds have been committed.

However, each phase of Senior Trend only started after certain key events in the Have Blue program had progressively validated the concept of a faceted, very low observable aircraft. Formal Air Force requirements studies of a low observable strike aircraft were begun (August 1976) only after the Have Blue "pole-off" had been completed (April 1976), validating the potential of new shaping techniques to dramatically reduce RCS. As the demonstrator design was refined during phase II of the Have Blue program (1976–1977), and progressively higher fidelity aerodynamic and RCS model test data was generated, the RDPJ staff kept apprised of these developments and concurrently refined the requirements for an operational very low observable strike aircraft. The formal ATA concept definition contract was not awarded until October 1977 when the Have Blue design, development, and fabrication were essentially complete and only two months prior to its first flight. By the time the F-117A full-scale development contract was awarded in late 1978, complete flight performance and initial full-scale, in-flight RCS results were available from Have Blue (Have Blue No. 2, the signature test aircraft, began flying in July 1978). All Have Blue flying had been completed by November 1979 when the first F-117A long-lead production contract was approved.

Although production was well under way before the first Senior Trend full-scale development flight in June 1981, program managers recognized and accepted that some production redesign and retrofits would be necessary. In fact, a provision was included in the first lot production contract that required Lockheed ADP to provide modification kits to correct all problems found during initial full-scale development flight testing.[181] Furthermore, the low production rate minimized the cost and schedule impact of the various changes that resulted from the overlap between testing and production.

The increased procurement from one squadron to roughly a wing was proposed, and prior to first flight long-lead funding provided for a smooth transition from the original 20 aircraft into the extended production run. However, full funding of the increased procurement was conditional upon successful in-flight signature and avionics tests.[22, 92] Even after these test milestones were accomplished, each fiscal year's lot buy underwent a separate decision process at the time it was funded. Thus, the major decisions in the Senior Trend program were based on the successful accomplishment of key risk reduction events each step of the way.

Advantages of a Concurrent Program

Concurrency allows quicker program execution and earlier initial operational capability. This can work synergistically with streamlined management to minimize bureaucratic buildup and outside interference, which can derail a program from achieving its original goals. A quick program is also less vulnerable to funding fluctuations, loss of support due to political events, and high overhead because the program duration is shorter. According to Paul

Kaminski, "A compressed acquisition schedule was key to the success" of the Senior Trend program.[115]

With an aggressive effort to achieve an early, if limited, initial operational capability, early operational experience can flow into the later stages of system development *as well as into related programs.* This was a particularly important consideration in Senior Trend, which represented the U.S. Air Force's first opportunity to field a very low observables system; but it can be beneficial in any development program.

Team continuity is easier to maintain when there is some overlap between the phases of a program; experienced people are less likely to "drift off" to other projects. Many of the key people, both at Lockheed and in the Air Force, were involved from the early stages of Have Blue until the F-117A achieved initial operational capability. Lockheed ADP staffing and hiring policy reinforced this continuity and encouraged individuals to take "cradle-to-grave" responsibility for their work. Designers consequently gained more experience with production, flight test, operations, and support, thereby elevating the awareness of these activities in the design community as a whole.

The opposite extreme from concurrency is to have gaps between the stages of development. If this occurs, experienced people will be lost. An example is the Lockheed Jetstar; according to Kelly Johnson: "There was too great a gap in time between the first two Skunk Works prototypes and the production airplanes, so lots of benefits of prototyping were lost." [106] These benefits include rapid risk reduction and developing a knowledgeable team of people who "know" the airplane. The Skunk Works was a very "individual-oriented" organization, achieving rapid and cost-effective results by using small teams of highly capable individuals, but by the same token being more severely affected by the loss of those individuals than a larger organization would have been. In fact, it was not unusual at the Skunk Works for key individuals to be called out of retirement when their expertise or experience was needed. For example, when the TR-1 (a development from the U-2) was developed, certain engineers from the original U-2 project were rehired as consultants.[194]

Special Requirements for Concurrent Programs

Compressed, concurrent programs do require more flexibility than conventional programs. The original schedules for Senior Trend were very optimistic, and there were slips. When original schedules are aggressive, delays will sometimes have to be *accepted*, but should never be *directed*. Flexibility should be maintained to trade off nonessential requirements where appropriate to achieve payoffs in other areas or to protect the schedule. Mechanisms should be provided to address unforeseen problems For example, freedom to transfer funds between procurement and research, development, test, and evaluation categories was a valuable tool in the Senior Trend program.

An interesting and more recent example of the same principle in the commercial world concerns developing the Boeing 777. When difficulties were encountered, Boeing executive Alan Mulally said, "We could have stretched it out but it just seemed best to us to keep the end date the same and add some more resources."[27] According to an article in *Acquisition Review Quarterly,* "The wisdom of Mulally's decision was proven a thousand times over.... [Boeing] was committed to a successful 777.... As a result, Boeing management and its Board of Directors were focused on what they had to do to make it all happen. They were willing to commit Boeing resources toward overcoming potential challenges...."[27] The approach of stepping up the short-term effort to complete the job on time is exactly the opposite of what often happens when problems arise in DoD programs: the schedule typically moves to the right, which adds cost in the long run.

In a concurrent program some design changes will have to be retrofitted to production aircraft. The Skunk Works' procedures for incorporating changes were to integrate the changes onto new production airplanes as rapidly as possible to accommodate the needs of concurrent development and production. According to Lt Col Randy Smith, formerly of the 2762nd Logistics Squadron (Special), Norton Air Force Base, which supported both the SR-71 and the F-117A,

> ECPs [engineering change proposals] were worked in an ambitious manner without the need for government approval of detailed drawings. Parts were on the factory floor as soon as an ECP was approved for SR-71. I suspect ATA was identical.... [Drawings] were released to the shop floor ASAP, even before systems were fully designed (to reduce lead time on critical parts).[216]

To track changes, the Skunk Works did have policies for changing the drawings, resulting in a very well controlled system in spite of the aggressive pace. Even when handled efficiently, however, incorporating design changes will result in many individual airframes having their own unique configurations.

Many capabilities for the F-117A were not addressed until after limited initial operational capability. Automated mission planning, weapon delivery accuracy, flight envelope limitations, and reliability and maintainability, were all outstanding issues at that time. However, during the period following initial operational capability the issues were all worked, and the effort benefited from close interaction with the operational squadrons that were flying the aircraft. This interaction does not occur if initial operational capability is postponed until after all of the testing is complete. It requires a different attitude, but it can work.

Even in the most conservative acquisition programs, continued development and upgrades will be necessary after initial operational capability. This is because technology advances on a time scale that is becoming much more rapid than the typical weapon system development cycle. Advantages to relying on a certain amount of preplanned product improvement rather than trying to "perfect everything" prior to initial operational capability includes basing improvements on operational experience. Another advantage is that some of the necessary technologies will be more mature if introduced later and will often surpass the capabilities that can be built into the system initially. In contrast, perfecting everything on the very first production units will slow the pace of a program at a stage when the manpower and infrastructure—and therefore the spending rate—is high, driving costs up and increasing the program's exposure to shifting political priorities, budget cuts, changing requirements, and other external factors.

The F-117A had many limitations at initial operational capability in October 1983, but deficiencies have gradually been corrected and improved capabilities added. By August 1984, Lockheed was on contract for a variety of retrofits and modifications to the aircraft. Composite rudders were retrofitted to eliminate flutter problems, which restricted its maximum speed. Weapons control computers were improved, and weapons bay design improvements allowed the two bombs to be dropped simultaneously (except for laser guided GBU-27s, which required the weapon bay doors to open slightly beyond 90 deg to accommodate the relatively large tail fins). An improved mission computer allows "four-dimensional navigation," or precise control of time as well as position at each navigational waypoint. The cockpit was upgraded with color multifunction displays, a digital moving map, and a radar warning system. The landing gear was upgraded with new wheels and carbon brakes. The carbon brakes were not mature at the time the F-117A was in development but were a low-risk addition by the time they were incorporated on the F-117A. Technology is always advancing, particularly in computers and avionics, and low production rates are becoming the rule rather than the exception, this approach of gradual upgrades was very sensible.

LOW OBSERVABLES TECHNOLOGY (CHAPTERS 1 AND 2; APPENDICES A AND B)

Radar cross section (RCS) reduction techniques include the three broad categories of aircraft shaping, application of radar absorbent coatings (RAM), and special designs and treatments of details (such as antennas, canopy, access panels and doors, control surfaces, inlets, exhausts, apertures, and so on). Extensive and rigorous effort in all three of these areas is necessary to reduce detectability by a tactically significant amount. This cannot generally be accomplished through modification or retrofit to an existing aircraft design.

Radar detection range depends on many factors, most of which are under the control of the radar system designer, not the airplane designer. RCS, the only factor under the airplane designer's control, exerts only a weak influence on detection range. Specifically, detection range is proportional to RCS to the one-fourth power. To achieve a 50% reduction in the range at which a radar can detect an aircraft, RCS must be reduced to 1/16th of its original value (this corresponds to a 12 dB reduction). To reduce detection range by a factor of 10, RCS must be reduced by a factor of 10,000, or 40 dB.

SHAPING

Shaping was one of the least-understood of these three areas before the Have Blue program. Aircraft designers had only limited understanding of electromagnetic scattering theory and no analytical tools to help them design low-RCS shapes. Nearly all low-RCS aircraft designs up to that time, therefore, embodied only basic intuitive shaping principals to eliminate specular reflections from surfaces at certain aspects, primarily nose-on and broadside.

An improved understanding of electromagnetic scattering theory and the development of computational prediction methods based on the theory were critical to Have Blue and F-117A design development. These factors elevated the state of the art from intuitive, hit-and-miss design, to an analytical approach and dramatically reduced the turnaround time for accomplishing design iterations.

U.S. intelligence community exploitation of Soviet open source scientific literature provided important technical insights and significantly aided U.S. development of RCS predictive computer models. However, the results of this Soviet basic research did not come into a vacuum: extensive work in the United States on signature reduction—specifically on radar cross section prediction techniques—had occurred already in the years preceding Have Blue. This knowledge base enabled rapid implementation of theoretical developments that had come from the Soviet Union.

Furthermore, as noted elsewhere, the use of experimental measurement remains a necessary complement to computational prediction in the design of low observable aircraft. Testing is necessary to validate predictions and to

measure the effects of phenomena that are not yet understood well enough to be predicted.

Radar Absorbent Materials (RAM)

RAM development had been underway since World War II. By the time the XST effort started, computer programs allowed designers to define multilayered coatings with specific thicknesses of different materials and accurately predict their performance.[26] Extremely high-performance, low-cost and lightweight RAM was already available for nonaircraft applications, such as on buildings in the vicinity of ground-based radar installations, to eliminate unwanted reflections from the buildings. In these applications, unlimited thickness is available, allowing coatings of several wavelengths' thick or multilayered with interfaces between the layers carefully tailored to attenuate radiation at the desired wavelengths. Also, the environment is relatively benign compared to an aircraft in flight.

Design of RAM for aircraft applications is more difficult. Thick coatings are not usually acceptable. Coatings that provide high absorption within a limited thickness are necessarily heavier. Performance is also usually restricted to a smaller range of wavelengths than is the case with thicker coatings. This is one of the chief reasons why RAM alone is not the answer to aircraft RCS reduction. This was learned through several unsuccessful projects prior to Have Blue. The lesson was not lost, and the two contractors who advanced to the ground test phase of the XST program both had a broader approach to RCS reduction.

Coatings also must be able to stay intact and attached to the aircraft surfaces during flight. Some parts of the aircraft experience high temperatures, which impose additional requirements on the materials and adhesives. These were the challenges that were dealt with during Have Blue and Senior Trend. The F-117A was the first operational aircraft to use RAM extensively. Attaching and maintaining the coatings was found to be very labor-intensive, for example "$750,000 in labor hours just to put the coatings on," according to Alan Brown.[34] Improvements subsequent to initial operational capability, chiefly the use of spray-on RAM rather than glue-on sheet RAM, have helped the situation significantly.

The main challenges today are in making RAM more producible, supportable, and maintainable. These challenges are the subject of many ongoing initiatives. Incorporating RAM into the structure (referred to as radar absorbent structure, or RAS) rather than as a "parasite" treatment is one of many possible approaches. Issues specific to the shipboard environment must also be addressed for U.S. carrier air wings to be equipped with low observable aircraft.

DETAIL TREATMENTS

In general, any part or subsystem of an airplane that exchanges matter and energy with the surrounding environment, including information, cannot be treated simply by covering it with RAM, because that would prevent it from accomplishing its primary function. Special design approaches and treatments are necessary.

Before the Have Blue program, the primary focus for detail treatments had been on the largest sources of backscatter, such as engine inlets, canopies, and radar/antenna installations.[26] Conceptually, effective solutions had already been identified for some of these areas, but achieving robust, practical implementations remained a challenge. For example, although internal weapons carriage was the obvious solution to the RCS problem created by external stores, extensive engineering effort was required to address air loads, acoustics, and release properties.

Furthermore, as the breakthrough in shaping techniques made it possible to achieve lower signatures than ever before contemplated, smaller details became important, such as the air data system, doors and access panels, refueling receptacle, and others. Particularly challenging were developing an operationally suitable low RCS air data system and installing the FLIR/DLIR targeting and designation system behind suitable apertures. Major efforts were required to achieve satisfactory solutions. All of these developments had not been attempted during the Have Blue demonstration program, which appropriately was focused only on demonstrating reduced signatures of the basic air vehicle.

Developing and, particularly, integrating the propulsion system was perhaps the greatest challenge of all. According to Jack Twigg, who had been Have Blue program manager and continued to be involved with low observables programs at ASD during the development of the F-117A: "Considering what was done wrong.... It was in the technical area of management and design of hot section engine exhaust parts. The lesson is to always let the propulsion team, not the airframer, be responsible for this design and development test."[234] Some of the problems might have been avoided by making GE responsible for exhaust system design from the beginning with Lockheed providing guidance in the unique requirements of low observables. Where any new propulsion technology or integration challenges are involved, strong and cooperative involvement from both the airframe and the propulsion contractors is important. Regardless of how it was managed, the propulsion installation represented a tremendous technical challenge, inevitably incurring a lot of trial and error before the system was perfected.

Propulsion integration remains a challenge as we strive to develop low observable aircraft with higher flight performance. Whether the application is a high-altitude, long-endurance drone or a supersonic fighter, the penalties associated with special low observable propulsion installations must be minimized. Durability of the exhaust system will remain an issue as low observable aircraft are developed with more powerful, afterburning engines. Hopefully the experiences with the exhaust system of the F-117A will not be forgotten.

OVERALL PROGRAM MANAGEMENT

Several program participants, representing most of the major organizations that were involved, have attributed much of the success of the early U.S. Air Force low observables effort to the senior leaders who structured the program, supported it, and allowed it to happen in a streamlined way with abbreviated oversight. They allowed much of the usual bureaucracy to be cut out and authorized program personnel to accomplish their jobs. Streamlined execution is impossible unless those in top leadership positions are willing to relinquish some of the traditional oversight and control.[7, 25, 135, 181, 234]

Streamlined program management assisted in achieving rapid results. The Senior Trend program followed established guidelines for special management. Every aspect of the program was characterized by small staffs reporting directly to key decision makers. The low observables project office in the Pentagon initially consisted of only five officers and had access to top Air Force and DoD officials and members of the applicable Congressional committees (HAC, SAC, HASC, and SASC). Similarly, the Aeronautical Systems Division (ASD) staff was initially less than a dozen people (part of a larger "basket" organization) and reported directly to the ASD commander. The 4450th Tactical Group was also small, streamlined, and operated through an abbreviated chain of command.

Even as the Senior Trend program proceeded through full-scale development and production, and other programs were added to the low observables effort, the ASD Classified Aeronautical Systems Program Office remained small. The staff peaked at 133 in the early summer of 1984, immediately before the Advanced Technology Bomber (B-2) split off into an independent office. After the B-2 split off, only 68 people were left to manage Senior Trend and at least two other classified programs, plus the low observable technology base, which was treated as a program in itself.[35] For comparison, Systems Program Office staffs typically have ranged from 250 to 550 during critical phases of programs such as the F-15, F-16, C-17, and F-22.[214] Recent Air Force acquisition streamlining initiatives are aggressively attempting to reduce the size of the program office staff needed to execute major programs.

However, each member of a small program office staff must be highly qualified and multiskilled, as well as possess personal qualities of dedication, openmindedness, ability to think on their feet, and the courage to communicate bad news as well as good news. Only handpicked personnel with unusually high levels of experience and professional competence were assigned to both the F-117A System Program Office and the Combined Test Force.* This philosophy was carried over from the Have Blue program.

Additionally, administrative burdens must be kept to a minimum, otherwise a small program office will be completely overwhelmed. Most program offices expend a lot of time preparing briefings, working budget "what if" exercises, and responding to inquiries both from critics and supporters of the program. The F-117A SPO was generally protected from such burdens primarily by three factors. First, program secrecy kept outside inquiries and other "busy work" to a minimum. Second, whenever possible, the RDPJ office handled program coordination activities in Washington without placing unnecessary demands on the SPO.[116] Finally, contracting and decision authority were simplified and streamlined as much as possible consistent with statutory regulations. Coordination and oversight were accomplished in "near-real-time" on practically a face-to-face basis. All key players had an intimate knowledge of their roles on the program, of one another, and of overall program issues and challenges.

To execute a program in an abbreviated management environment, program personnel must know clearly what they are trying to accomplish. Each phase of the F-117A's development, from the initial DARPA studies, to the Have Blue demonstration program, through limited initial operational capability, had a small set of clear, highly focused goals and objectives. In engineering, documentation requirements were relaxed relative to a traditional program:

> Traditional process requires government experts in all product areas to provide specs for every aspect of a system.... Special Management process requires trust in contractor's technical ability and government's commitment.... [Government] experts are typically called on only as required, i.e., management by exception versus full-time oversight.[216]
>
> Perhaps this is the one area that offers the most promise of reducing overhead...the requirement for documentation that will rarely, if ever, be used is no longer affordable. Government engineering oversight by numerous individuals within a particular discipline needs serious review.[87]

On the F-117A program, "Government oversight was provided by each function to ensure a necessary system of checks and balances. However, each

* In the case of the CTF, "All pilots, both Lockheed and U.S. Air Force, on the RDT&E side were highly qualified, had many flight test hours, and were graduates of the U.S. Air Force Test Pilot School. All had flown many types of air superiority and ground attack aircraft and knew what it would take to made this aircraft great. All the U.S. Air Force flight test engineers were experienced graduates of the U.S. Air Force Test Pilot School Flight Test Engineer Course and had vast amounts of experience in setting up and implementing test programs under hazardous conditions with little or no supervision. All CTF personnel were go-getters who could be trusted to get the job done in a minimally manned environment."[212]

function was provided by the *minimum* number of people required to do the job. Engineering oversight was provided by *one engineer for each discipline* (aerodynamics, propulsion, structures, low observables, and others). One government contracting officer and one auditor performed their functions. As seen in Fig. 15, actually a handful of people in each discipline were officially "shared" among several programs; but in practice each program would rely primarily on a single individual within each discipline. The cognizant government overseer decided when to seek outside advice without flooding the contractor with unwanted and often counterproductive "help." In contrast, a typical major weapon system today will have in excess of 100 engineers providing oversight."[87]

Communicate, communicate, communicate was instilled and worked in the Have Blue and subsequent early U.S. Air Force low observables programs.[234] Within the Skunk Works, engineers in different disciplines had quick access to each other, the people on the shop floor, and the Skunk Works manager when they experienced problems, following the principal that "bad news must travel to the top at least as quickly as good news."[197] On the government side, the small low observables office in the Pentagon reported directly to key decision makers and maintained an open, nonadversarial relationship with Congress. The RDPJ staff was "brutally honest" with Congressional committee members and thereby gained their trust.[115] This helped to assure stable funding for the program. Although people have a natural reluctance to report bad news, hoping perhaps that problems will "go away," inevitably the consequences are worse when the problems surface later.

Because of the covert nature of the program, rapid and open communication depended on having "carved-out" or dedicated secure voice and data transmission capabilities at all involved critical program sites. "The ability to work matters *real-time* and totally secure saved untold man-hours, man-days, man-years.... The DACOM net was one of the largest in U.S. government, one of the program's major achievements," according to former Senior Trend Security Manager Peter Knauth.[121]

An important aspect of the Classified Aeronautical Systems Program Office was that it managed several programs besides Senior Trend. According to Lynn Brucker, an avionics engineer at the SPO, "The large number of programs in different phases of development [Advanced Technology Aircraft (F-117A), Advanced Technology Bomber (B-2), Tacit Blue, and others] resulted in a very broad experience base for the SPO personnel."[35] This also provided a balanced perspective. Personnel in a single-program office may feel threatened by the possibility of the failure of that program, whereas an office that manages multiple programs can be more objective, learning from those programs that fail, and then concentrating its efforts on the more promising ones.

Focus on developing one or more specific weapon systems can take attention and funding away from equally important technology activities or lead to "tunnel vision" regarding a system's potential weaknesses and alternative approaches. This was not allowed to happen to the Air Force's early low observables efforts. In August 1981 then-Col Paul Kaminski initiated action to maintain the necessary broad perspective and continuing attention to the technology base:

> The [low observables] program emphasizes expedited development and introduction of low observables platforms into our operational forces. The program is maturing, with...Senior Trend...transitioning into production by FY83. This transition will draw attention and resources away from broadening the LO [low observables] technology base. The recommended addition [of 1% of the total FY83 program funding] will provide for a dedicated "Red Team" to better understand the vulnerability of low observable vehicles, a "Blue Team" effort to exploit these vulnerabilities in developing defensive reactions to LO technology, and a continuing program to broaden our current technology base for future offensive applications.[110]

This was one of many foresighted decisions by early planners, which have led to the current U.S. technological lead in low observables.

Another important factor was the empowerment of highly motivated, working-level people, including many relatively junior officers. Jack Twigg recalled, "I had dedicated and expert people and had the full *authority* and *charter* to operate and succeed" on the Have Blue program.[234] During the concept exploration and requirements formulation that led to Senior Trend, the RDPJ office was staffed by one colonel and four majors. Relatively junior officers similarly played key roles in establishing the security structure at RDPJ and at the SPO. Many of the people from all of the various organizations stayed with the program as they were promoted, providing informed and experienced leadership as the program reached its critical stages. During testing, working-level people from the SPO, the Air Force Flight Test Center detachment, the Tactical Air Command operational unit, and the contractor worked together in a largely nonadversarial environment and with the authority to perform their jobs. This interaction played a key role in solving problems and accomplishing objectives, for example in developing the GBU-27 precision guided munition.[170] According to then-Col James Allen,

> Every pilot, maintenance officer, and many of the senior maintenance technicians were not only assigned to develop/build part of the unit operation (facility design/procedures/policies, etc.) but each was also assigned an aircraft subsystem to work. Perhaps most important of all is that each had decision authority within their area; they did not have to

> come back to me as the commander for approval. And in nearly every case the people rose to the challenge.[7]

Many of these considerations can be summarized by noting the importance of three qualities that, according to several sources, made the "black world" work: talent, trust, and integrity. All of the organizations involved were staffed with capable, motivated personnel. There was trust between and within the different organizations, so that work did not have to be double-checked or duplicated by others. The trust was backed up with integrity; the government and the contractor did not conceal information from each other, and mistakes and setbacks were promptly and honestly communicated.[8, 135, 235]

Finally, the early low observables effort benefited from certain favorable circumstances that were beyond the control of any individual or group. The low observables program enjoyed "strong consensus among the senior leadership of the Air Force, DoD, and the bipartisan leadership of the four applicable Congressional oversight committees that there was a requirement to rapidly introduce this revolutionary capability into the combat force...." [87] This was in part a result of careful planning and advocacy-building by those who were involved very early in the program. But it was also a result of outside factors: "requirements pull" (Vietnam and 1973 Middle East experiences) combined with "technology push" of *revolutionary* magnitude (a series of significant developments in RCS prediction techniques, enabling a novel and aggressive design approach that was validated first in pole model tests and eventually in the Have Blue flight demonstrations).[115]

Such technical breakthroughs are a rare occurrence, and when coupled with an operational need, offered an almost unique opportunity to run a program with strong support at the highest levels. Support was further reinforced by the early 1980s Reagan Administration military buildup, which began just as the F-117A was entering its critical flight test phase. The support enjoyed by the F-117A is difficult to find in today's national security environment, which is increasingly cost-conscious and risk-averse. A modern program manager can be sure that the need for any new weapon system will be questioned and that the risks associated with its development will be criticized. The high level of secrecy in the Senior Trend program had the side effect of insulating it from much of this kind of criticism. All of these factors would not have made the F-117A program successful without good management strategies and execution. But powerful outside factors worked in favor of the program.

Chapter 8

Covert Program: Favorable and Unfavorable Impacts

Developing the F-117 from phase II of Have Blue until several years after initial operational capability was completely covert. Not even its existence was acknowledged. The reasons for this were to protect the existence and nature of the program, the level of government commitment to the program, and the technology details.[121] The success of the Senior Trend security effort in accomplishing these goals is beyond question.

Streamlined, flexible management approaches have been successfully used in both "black" and "white" programs, but have usually only worked when the highest levels of authority, both in the government and in the participating company, have been committed to assuring success, often bypassing traditional practices and procedures.

Advantages of a Covert Program

Undoubtedly, certain aspects of the tight program security worked in favor of the F-117A development. A number of technical difficulties and shortfalls encountered during full-scale development would have caused a lot of unwanted attention if the program were more widely known. Both Have Blue demonstrator aircraft and the first F117-A production aircraft were destroyed in crashes. Initial operational capability was delayed. The flight envelope at initial operational capability was restricted in speed, altitude, and *g* capability (both positive and negative) relative to initial goals and predictions (by 0.1 Mach, 5000 ft altitude, and $\pm$ 2 *g*, respectively.[179] Weapons employment

capability was also less than planned. The small number of people involved realized that these shortfalls could be accommodated and eventually overcome, and the F-117A still offered a fundamental and unique mission capability. It needed to be operational to gain experience with the whole new discipline of low observables. The media, Congress, and others involved are often not so understanding. Senior officials might have micromanaged the development when technical troubles arose. Congress might have stretched the program to reduce concurrency, thereby increasing costs. The crashes or the initial performance shortfalls might have fueled political opposition to the program, possibly culminating in cancellation. Instead, protected by secrecy from outside attention and intervention, the Air Force and the contractor worked together to solve problems, accepted delays when they could not be avoided, and traded off performance when it made sense to do so.

The highly secret environment probably improved the willingness of people in the program to promptly report problems up the chain of command. This willingness facilitated the rapid, cooperative approach to problem solving that characterized the F-117A program. However, this has clearly not been the case on every "black" program, as evidenced by the massive and catastrophic failure to report problems while developing the A-12 (the "new" A-12, not the Lockheed A-12 of 1962). A willingness to promptly report problems is an attitude, and secrecy is neither the only way nor probably the best way to foster this attitude, but it worked in the case of the Senior Trend program.

Tight program security and streamlined management, both of which were elements of the special management regulations that governed the low observables effort, worked to keep the program small and focused. The ATA project was initially managed by a staff of five officers hidden away in the strategic reconnaissance office in the Pentagon. This was exactly the type of organization that the Lockheed Skunk Works and similar organizations liked. Then-Brig Gen Lawrence Welch, who was in charge of planning at Tactical Air Command headquarters and later Air Force chief of staff (from 1986 to 1990), became the chief operating officer for the Air Force low observables effort, and met once a month with William Perry to make important decisions on the program. Gen Welch noted, "Sometimes he agreed, sometimes not, but we never had delays.... Because we were so highly classified, the bureaucracy was cut out and that made a tremendous difference." [197]

The F-117A's covert status also kept it out of *visible* competition with other programs, and out of the political arena in general. "The fact that the F-117 was a highly secret program insulated it from the types of organizational and bureaucratic maneuverings that characterize the development of other American fighters. The secrecy and the fact that the program was relatively small, permitted the contractor to take technological risks and hence develop capabilities that might otherwise not have been possible." [236]

Secrecy provided protection from distractions, which can have a crippling effect on the productivity of even the most capable personnel. An increasing problem in DoD programs is the time spent responding to government inquiries, answering questions, and justifying every activity or decision. Even the most ignorant and specious criticism must be answered or it becomes a rallying point for the opponents of a program. All of this takes away from the time available for real work.

The existence of the F-117A was acknowledged in 1988 at which time a few deliberately poor-quality photos were released. In April 1990 the F-117A itself was finally revealed to the public. By that time, the F-117A was a mature system, as illustrated by its strong performance in the Gulf War less than a year later. "Next-generation" low observable aircraft such as the B-2 and the Advanced Tactical Fighter were already well into development. According to former F-117A Security Manager Lt Col Peter Knauth, U.S. Air Force (Ret),

> One of the key secrets we tried to protect during the F-117 FSED [full-scale engineering development] program and follow-on production was the scope and level of commitment being devoted to the weapons system. In other words, not so much *what* was involved (stealth technology) but the real objective of getting VLO [very low observable technology] on many militarily useful aircraft as fast as possible. I have heard (and believe) that at one point we would want the "bad guys" to know what we were doing and how serious we were about doing it so that, as with Star Wars [the Strategic Defense Initiative] the Soviets would become convinced that the U.S. would do whatever was required.... These revelations, plus the unstoppable quantum leaps in the worldwide exchange of general information, made the Soviets realize they could no longer tax their people to beat us or isolate them behind walls. Hence the 1990s developments, i.e., the collapse of the Communist Soviet government.[121]

Possible Drawbacks of a Covert Program

A danger of a covert program is that there is less opportunity for independent analysis of the system being developed. A situation can arise where the only people who know about a program are its advocates, and weaknesses in the system or its concept of employment can be overlooked. This effect has probably contributed to some serious problems in other "black" programs. To avoid such a situation, Col Paul Kaminski set up a formal counterstealth program, or "Red Team," shortly after he became the director of the low observables office at the Pentagon in March 1981.[110, 115]

Few military planners knew about the F-117A, and even fewer had any real confidence in its capabilities. A significant effort was needed to match stealth technology to employment doctrine, addressing the question of whether it was simply a "silver bullet" for special operations, or a "force multiplier" that could, for example, allow conventional aircraft greater freedom to operate by disabling enemy air defense systems. Some people were uncomfortable with the concepts of flying only at night or of a "fighter" with no air-to-air armament. On the other hand, some had the idea that the F-117A was invisible and invincible. Modeling and simulation played a key role in defining how the F-117A could operate and what threats or situations it would have to avoid. The experiences of Desert Storm proved out the conclusions of this effort.[115]

Considering the cost of the program, the security burden during the development stages was probably most pronounced on the contractor. Ben Rich estimated that the tight security in the F-117 program increased costs by 10% to 15%, from the contractor's perspective.[197] At the SPO, estimates are that the covert program cost 5% more than *the same program* would have cost in the "white world," remembering that every program has technology and performance data that must be protected.[163] However, those figures must be weighed against the fact that conventional programs often take 15 years or more from go-ahead to initial operational capability with SPO staffs growing to several hundred; so it may be reasonably estimated that the covert status of the program actually saved money during development by helping to keep the program small and quick.

Probably the biggest added cost was operations. A new operational base had to be established at Tonopah at a cost of significantly more than $100 million. Permanent housing for F-117A personnel and their families was not at Tonopah but at Nellis Air Force Base, so the Air Force transported about 2300 people weekly between Nellis and Tonopah using leased commercial aircraft. All F-117As were transported in C-5s between Lockheed Burbank and Tonopah, and between Tonopah and the F-117A depot in Palmdale. C-5 movements were done at night only; the F-117A's wings had to be removed for the C-5 moves. Another consequence of flying a nonacknowledged airplane was the need for a "surrogate" for training and other purposes. Approximately 20 A-7Ds were therefore operated by the 4450th Tactical Group. The cost of all of these special provisions can be avoided in "gray," acknowledged programs where concealment of a program or prevention of all sightings of a system by the public is not essential.

The public acknowledgment of the F-117A in 1988 and the declassification of many details allowed many of the special security procedures of F-117A operations to be relaxed. A July 1991 report by the Air Force's Sacramento Air Logistics Center's Directorate of Specialized Management noted that there would be a shift to routine logistics management of *many* airframe, engine, and systems requirements, but that "limited technical data, a small fleet, and high-cost reprocurement drawings make competitive reprocurement or organic repair impractical for most program unique components."[47] Continuation of special management was therefore recommended. When the Air Force decided not to obtain maintenance manuals and technical data early on, it had basically committed the F-117A to "long-term contractor relationships in engineering, logistics, and test support."[47] Some advantages and some drawbacks are present in this kind of relationship. Favorable aspects are that the contractor develops experience and a thorough knowledge of the system, can concentrate on getting the real work done instead of preparing proposals, is more likely to report problems honestly and promptly, and is more likely to make long-term investments to provide improved products or services if the company knows that the program will not be moved to its competitor in the near future. In short, trust is established. The chief drawback is that the customer has very little ability to change the way things are done. Specifically, due to the increased procurement, the Air Force desired to assume a greater part of the F-117A maintenance burden, but there was little incentive for Lockheed to make this easier for the Air Force. The Air Force's dependence on contractor maintenance made achieving badly needed reliability and maintainability improvements very difficult.

One consequence of "normalizing" certain security aspects of the F-117A program is that "agencies involved in weapon system management and funding authorizations are beginning to ask for the detailed information normally required...." including planning documents, technical data, and financial information. Much of the documentation in question was never generated and would be impossible or at least very expensive to reconstruct now.[47] Oversight organizations do not like to be excluded from programs, and so they will always resist the application of streamlined management. All programs, whether "black," "gray," or "white," need to remain focused on their original goals, maintain a cooperative approach to problem solving, and keep nonvalue-added work to a minimum.

SUMMARY

In summary, key benefits of the covert nature of the F-117A program were that it did the following:

1) Protected the existence, level of effort, and technology details of the low observables program from exploitation, and allowed the United States to decide if and when to reveal the existence of the program and its technology.

2) Allowed the program to survive technical difficulties that normally lead to increased oversight, micromanagement, delays, or even cancellation.

3) Fostered an open and cooperative attitude among those who were part of the program.

4) Helped, in combination with other elements of special management, to keep the program small and focused on its original goals.

5) Kept the program out of visible competition with other programs and out of the political arena in general.

6) Provided freedom from outside distractions.

The most significant disadvantages were the following:

1) The potential for weaknesses and vulnerabilities could have been overlooked.

2) Few military planners were aware of the existence of the system, its capabilities, and its limitations.

3) There were added costs of reportedly 10 to 15% for the contractor and 5% for the program office relative to *the same program* (in other words the same duration and manpower) under conventional security.

4) There were added operating costs particularly due to nonacknowledged program status and resulting special basing and support requirements.

5) Difficulties were encountered later while attempting to normalize some aspects of maintenance and logistic support.

Chapter 9

Could It Happen Today? Changes in the Acquisition Environment

The Senior Trend program operated under a highly abbreviated oversight structure. This is not to say that it escaped oversight but rather that specific but abbreviated reporting procedures were associated with Special Management. Key officials up to the very highest levels of Air Force, DoD, and civilian government leadership were kept extremely well informed about the program.

Government oversight has, however, become more complex since the F-117A was developed. A more detailed system is in place to ensure that a need exists, a new weapon system is in fact necessary to meet that need, requirements are valid, and effort is not duplicated with other services before a new weapon system acquisition is begun. Select members of the various acquisition boards and committees are cleared into special access programs to perform the functions of those committees in the "black" world. Thus even a covert program today may not be able to take some of the shortcuts that were possible 20 years ago. The key then to timely success in "black," "white,"

and "gray" programs is to develop the methods for steering them through the acquisition process with a minimum of unnecessary regulation, interference, and delay. This is a challenging task, and the acquisition environment is continually changing:

> Acquisition management has been in a constant state of flux. One significant observation over the past 50 years is the cyclical nature of moving from cost to fixed price contracting, centralized control to decentralized control, and separated commands to unified commands for acquisition and support. This cyclical nature of change with the same problems persisting begs the question of whether or not we have really identified the root causes of the problems, or if we just get fed up with the problems and try to change things to see if the problems will go away.[192]

Hardware Demonstration and Risk Management

The modern economic and political environment requires that developers of new weapon systems manage risk and be very conscious of cost. However, it can be counterproductive to "over-manage" risk, according to Paul Kaminski, under secretary of defense for acquisition and technology.[115] Micromanagement, excessive regulation, and requirements for overly detailed documentation and justification only interfere with real work and tend to *discourage* open communication of problems lest it precipitate still more interference. Tight budgets are not new, but present prevailing responses to that circumstance appears to be a desire to ensure that projects will never fail—an almost paranoid aversion to technical risk.

Along the same lines, concurrency has tended to become a bad word. However, the F-117A's compressed acquisition schedule has been called a "key to its success."[115] Basically, a good design or a sound concept under capable management can often be successfully developed under a concurrent program. A bad design should never be developed at all. Early hardware demonstration efforts, such as Have Blue, are often the best way to distinguish between the two and may be the key to *managing*, as opposed to *avoiding*, technical risk. Once the key risk issues have been addressed through such demonstrations, a program may proceed in a rapid and streamlined manner, as the Senior Trend program did, with confidence that is based on proven concepts and technologies rather than elaborate oversight. Obviously, risks remained after Have Blue, but the fundamental concept had been validated. The remaining risks were in specific areas such as avionics, apertures, propulsion integration, and so on. Risk also was present and in the overall design refinement for an operationally suitable weapon system. All of these risks are appropriate to the full-scale development (or in the current terminology engineering and manufacturing development) phase of a program.

Taking some risk is essential to providing advances in military capabilities, and an excessively risk-averse attitude will stifle innovation and thereby miss out on cost-effective breakthroughs. This forces the military industrial complex into a plodding brute-force approach to technology, which can be more costly in the long run than a more adventurous approach where some programs fail but others make significant steps forward. The defense acquisition environment, therefore, needs to accommodate risk by minimizing the costs of failure, rather than avoiding risk altogether.

The early 1970s was the period of the highly successful Lightweight Fighter program involving the YF-16 and YF-17, from which the F-16 and F/A-18 were eventually developed. Such demonstrator vehicles were referred to by the highly appropriate term experimental prototypes. These demonstrators or experimental prototypes could, as appropriate, be subscale or considerably simpler than the conceived operational weapon system. As a result, the costs can be lower and the consequences of failure for a small demonstration effort much less than for a larger scale demonstration or high- fidelity prototype effort conducted under a major acquisition program.

The whole definition of acquisition has been refined and broadened and is currently recognized to include all phases of a weapon system's life cycle from initial concept exploration to retirement and disposal. While this promotes a more holistic, interdisciplinary, "big picture" approach to weapons system development, it may also impede early "experimental" prototyping and hardware demonstration and accentuate the more cumbersome aspects of the acquisition process.

The conventional acquisition cycle (illustrated in Fig. 26) begins with identifying a need and preparing a mission needs statement. For a major program, such as a new aircraft, this is reviewed by the Joint Requirements Oversight Council (JROC) and then submitted to the Defense Acquisition Board (DAB) for approval to begin concept exploration and definition, or "phase 0." During this initial phase, alternatives are considered, study contracts awarded to industry, and an acquisition strategy developed that addresses a wide variety

	MS 0	MS I	MS II	MS III	
	Phase 0	Phase I	Phase II	Phase III	Phase IV
(Determination of Mission Need)	Concept Exploration & Definition (CE/D)	Program Definition & Risk Reduction (PD&RR) (Formerly Dem/Val)	Engineering & Manufacturing Development (EMD) (formerly FSD)	Production & Deployment (Prod/Depl)	Operation & Support (O&S)

Fig. 26 Traditional acquisition process through phase IV. [137]

of topics (such as free and open competition, contractor past performance, environmental impacts, live fire testing, and long-term production considerations). Several documents must be prepared, including the initial versions of the operational requirements document, cost and operational effectiveness analysis, test and evaluation master plan, and system threat analysis report. An acquisition program baseline is established. At the completion of concept exploration and definition, an integrated program summary is prepared and submitted with supporting documents to the JROC and the DAB for approval to enter phase I, program definition and risk reduction (PD&RR).[192]

In addition to conducting technology demonstration activities in phase I, the program office must update existing acquisition documents and create documents related to phase II, engineering and manufacturing development. Although acquisition reform initiatives are being implemented to simplify the process, and abbreviating the list of documentation requirements as appropriate to a particular program is encouraged, a large number of requirements generally must still be satisfied before any hardware is built and tested.

By this time, many people who have advocated the system, performed the analyses, and prepared the necessary documents are now part of a program office and have a vested interest in the program's success. A bureaucratic, risk-averse environment may therefore be created before significant hardware demonstration can even begin. This is not to say that the various documents and plans do not serve a valid purpose; they are an important system of checks and balances, helping to ensure that risk is progressively reduced as a program matures with no unpleasant "surprises" later in the program. However, attempting to eliminate too much risk too soon will often stifle innovation. When aggressive, new technology is involved, placing some of the more formal elements of the acquisition process after the key technology demonstrations, not before, may be more appropriate.

Have Blue was a technology demonstration program, not an acquisition program, there was no attempt to put a whole acquisition strategy in place until the airplane was built, had validated the technology, and had provided insight into what sort of operational low observable aircraft would be technically feasible. The Have Blue program effectively served as an informal demonstration and valuation phase for the F-117A. The key difference was that it started *before* formal Air Force studies of an operational very low observable aircraft. The Have Blue/Senior Trend acquisition cycle is illustrated schematically in Fig. 27.

Several benefits resulted from this approach. The Have Blue project office and contractors could concentrate totally on the technical issues, not on meeting the requirements of a formal acquisition program. The F-117A concept exploration phase and requirements formulation process could benefit from the results of Have Blue demonstrations. If Have Blue had failed, relatively little money would have been wasted. But its success, and the subsequent rapid application of the technology to the F-117A, gave the United States a unique asset and a 15- to 20-year (or more) lead in low observables technology. Perhaps the most important lesson to be learned from developing the F-117A is that demonstration and valuation-like activities—technology demonstration or "experimental prototyping"—of aircraft with a high degree of relevance to military needs, should in some cases be allowed before establishing a formal acquisition program.

This is not a new idea but rather has surfaced periodically throughout the history of modern military systems acquisition. In fact, this was the strategy that was successfully employed to advance the state of the art through the first several decades of military aviation, even during the extremely tight financial situation of the 1930s. Defense contractors develop advanced systems, sometimes with their own funding, and demonstrate them to the military in the hope of obtaining orders. Later, in the early 1970s, Deputy Secretary of Defense David Packard, backed by a Presidential Blue Ribbon Panel, revived the idea of an experimental prototyping effort with the intention of starting two or three demonstration programs each year without ties to subsequent full-scale development or production.[13] Unfortunately, only a few programs started under this effort, most notably the Lightweight Fighter Program (YF-16 and YF-17). In the "black world," where even the existence of a program is not allowed to be known, formal Special Management guidelines allowed certain high-priority programs to bypass some of the conventional acquisition procedures, and the low observables programs were conducted under those guidelines. Today, advanced concept technology demonstrations fit the same model of streamlined demonstration effort, which are not tied to the requirements of a formal acquisition program. Advanced concept technology programs can serve both to validate and mature new technologies through flight testing without creating major new programs. They are a way to stay on the leading edge without excessive expenditures or the difficulties of formally establishing and executing the requirements of major acquisition programs.

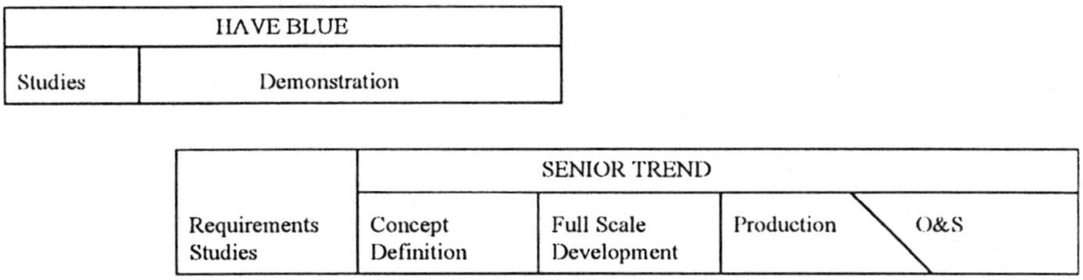

Fig. 27 Schematic diagram of Senior Trend acquisition phases.

To ensure that such prototyping efforts are meaningful and not just "hobby shop" projects, operational communities need to be included in the selection and management of prototype projects. Clear goals and objectives should be set so that results are not ambiguous but clearly demonstrate success, failure, or specific shortcomings that can be addressed in future work. Both of these considerations—military relevance and clear goals—were adequately addressed in the Have Blue/Senior Trend programs.

Other Considerations

Production Rates

Ever since World War II when military aircraft were produced by the thousands at a rate that used the full capacity of a mobilized U.S. industrial base, the trend has been toward procuring smaller numbers of aircraft at progressively lower rates. This is conducive to intelligently planned concurrency and the gradual upgrade strategy used in the F-117A program. With a low production rate, a smaller number of units are affected by any design changes. Rapidly introducing a system operationally, with some carefully chosen limitations and then upgrading it is probably more cost effective than trying to develop a fully mature system at initial operational capability. This may involve a 15+ year development effort! Combat aircraft undergo continual modifications and upgrades anyway because of advancing technology and changing operational needs.

Computer Technology and Electronic Communications

Computers and electronic communication are valuable tools for rapid decision making and for exchanging technical information. Dedicated, secure electronic communication played a key role in successfully executing the Senior Trend program. However, advances in information technology and the widespread use of networked personal computers that has occurred since the early days of Senior Trend are not necessarily a guarantee of increased efficiency or reduced total manpower. Computer networks and associated databases often require dedicated support teams, often provided by support contractors. Overuse of electronic mail, constantly updating massive files of briefing charts, and other types of unnecessary electronic busywork can be just as time consuming as "paperwork," and may actually be more expensive, while contributing just as little to the actual progress of the program. The key to improved productivity is to eliminate *all non-value-added* work, not necessarily just *paper*work.

Funding Flexibility

Senior Trend program management frequently exercised the ability to transfer funds from research, development, testing, and evaluation and procurement, and this allowed the program office to balance increased development effort to deal with unforeseen problems against the need to maintain a steady production schedule. A reform measure to allow transfers between research, development, testing, and evaluation and procurement budgets in conventional acquisition programs is reportedly under consideration at the present time.[200] Senior Trend experience indicates that this might be an excellent way to keep pace between production and development aspects of a program, particularly when there is some degree of concurrency.

CAN A SKUNK WORKS–TYPE ORGANIZATION OPERATE EFFECTIVELY TODAY?

A true Skunk Works–type organization is unquestionably more difficult now than it was 40 or even 20 years ago. It thrives on a relationship of independence and mutual trust with the customer and upper management of the "parent" company. Recent trends, both on the government side and within companies, have been for more oversight. This can inhibit creativity and reasoned risk-taking that characterize the Skunk Works approach. Some regulations must be relaxed. On the other hand, some of the newer regulations exist for good reasons and must be accommodated. For example, good environmental, health, and safety practices in production facilities simply need to become a way of life and need not stifle the creative Skunk Works spirit.

An increasing tendency for major weapon systems is to be developed and produced by geographically dispersed teams. This can occur as a result of industry teaming on major defense contracts or consolidation of previously independent companies into giant conglomerates. Either way, the challenge increases to apply a Skunk Works–type approach, which operates on real-time, open (often face-to-face) communication between coworkers, between workers and managers, and between designers and machinists. In spite of advances in information technology, an engineer at one company may find difficulty in talking to a machinist hundreds or thousands of miles away, even if the engineer knows exactly to whom to talk. Both individuals may not be able to simultaneously have their hands on the part, tool, or process they are discussing. At the Lockheed Skunk Works, designers would often visit the shop floor several times a day, form working relationships with the machinists, and find out how they worked. When a question arose, the people they needed would be there, not out at a meeting half way across the country. In a team effort today, not only is there geographical separation but institutional separation as well. Different companies may have slightly or significantly different ways of operating, as well as trade secrets that they do not want to share (a teaming partner on one program is often a competitor in another program). Some of these institutional barriers may persist for a time even after different companies are consolidated into a single conglomerate. A concerted effort is needed to ensure adequate technical and programmatic communication in these situations.

Communication across disciplines is equally important. Engineers are becoming increasingly specialized. While this is necessary to some extent because of the complexity and depth of available technology, it is also important to foster an appreciation and understanding of all disciplines involved in the design of a weapon system. Close contact between engineers of different specialties was an important characteristic of the organization that developed the F-117A. Elegant design solutions such as the U-2, SR-71, and the F-117A are not accomplished without a close-knit, interdisciplinary approach.

A Skunk Works approach is more challenging, but still possible to apply to modern weapons system development needs. Both customers and upper management of the "parent" companies must be willing to relinquish some control and allow a Skunk Works–type organization more than usual freedom to pursue innovative and cost-effective solutions to problems with a minimum of administrative burden. At the working level, government acquisition personnel must accept a more hands-off approach with less detailed oversight. In industry and government, the specialists who are the mainstream of today's engineering workforce must be strongly encouraged to learn about and participate in the activities of other disciplines. Bureaucratic barriers between all members of a team must be eliminated to facilitate problem-solving and allow the right technical decisions to be made.

CONCLUSION

The defense acquisition community as a whole recognizes that "acquisition reform will not be accomplished by adding regulations, but by removing those regulations that are impediments to effective decision making" and to streamlined, innovative design solutions.[115] Perhaps one of the greatest advantages to the streamlined management and abbreviated oversight that characterized the Senior Trend program was how it fostered effective decision making and problem solving by the working-level personnel who were best able to understand the real issues. As these people were promoted, they provided capable and experienced leadership in the later stages of the Senior Trend program and in other low observables programs. Many have since advanced to prominent leadership positions in the broader acquisition environment. They are now, therefore, able to apply their attitudes, experience, and lessons learned to other acquisition programs and to improving the defense acquisition process as a whole.

Perhaps the biggest task today is finding enough challenging work to maintain coherent teams of talented, experienced technical people. Along with lower production rates, new weapon systems are developed at progressively longer intervals. This trend offers fewer opportunities for engineers to gain experience designing real airplanes and other types of weapon systems. According to former Skunk Works President Ben Rich, "The old guard is retiring or being let go, while the younger generation of new workers lucky enough to hold aerospace jobs has too little to do to overcome a steep learning curve any time soon.... We are not producing enough airplanes for workers to learn from their mistakes."[197]

The Lockheed Skunk Works at the time of the Have Blue and Senior Trend programs had the ability to draw on experienced engineers from a wide variety of military aircraft projects including the YC-130, XF-104, U-2, A-12, SR-71, and S-3A. On the government side, the Senior Trend SPO was just one part of an office managing several classified programs, all in different stages of completion, and this environment provided SPO personnel with excellent perspective and experience.[35] Keeping design teams alive and providing experience for them would be an important benefit of sponsoring a larger number of technology demonstrations and experimental prototype programs such as Have Blue. More new project starts with relatively small initial commitments would yield more "successes" and would keep the industrial base and military test infrastructure not only funded but busy, which is just as important.

The low observables effort itself encompassed several programs besides Senior Trend, including the Advanced Technology Bomber and the recently acknowledged Tacit Blue reconnaissance aircraft. The promising efforts proceeded into full scale development, while others did not, but all of them provided valuable experience and yielded at least some contributions toward understanding of low observables technology, even if only how *not* to do things.

Kelly Johnson, founder of the Lockheed Skunk Works, established many of the principles that guided the development of the F-117A. He summarized the three most important lessons from 30 years of Skunk Works experience, all of which remain valid today.[106]

1) There is no single best method to develop an aircraft.

2) The most common cause of failure is lack of a suitable engine when needed.

3) The second most common cause is working to the wrong (or constantly changing) set of requirements.

According to Sherman Mullin, Lockheed's program manager for the F-117 from 1982 to 1985 and later president of the Skunk Works, "Aeronautical engineering was not meant to be a profession for those engineers obsessed with very low risk.... There is no such thing as a low-risk future program in advanced aeronautics. Yet this oxymoronic phrase is heard with increasing frequency. We must continue" to accept technical challenges and a certain amount of risk, "or we will slowly wither away." [180]

Innovation is vital to progress in transitioning new technologies into operational systems. If we become too risk-averse, we will stifle innovation and only very slowly will we develop new capabilities. While it is easy to say that the collapse of the Soviet Union alleviates the need for new military capabilities, one should look seriously at the environment in Russia and other former Soviet Republics: they are politically volatile and short on cash but not short on expertise or innovation. For example, a plethora of new Su-27 and MiG-29 variants have recently emerged along with very capable new air-to-air and surface-to-air missiles and advanced air-to-ground weapons. With third-world customers eager to buy systems that promise technological parity with the West, and the engineers at the former-Soviet design bureaus willing to apply an aggressive, innovative approach to improve prospects for foreign sales, we cannot dismiss the potential for significant advancement of threat capabilities. A corresponding need exists to pursue advances in our own capabilities. Furthermore, as our national defense posture adapts to a changing world, our forces will be called upon to perform different kinds of missions than in the past. We will need new types of capabilities to efficiently carry out our national defense priorities in the 21st century.

The F-117A development is not a pattern for every program but rather a useful example of how a unique technological opportunity can be quickly and effectively exploited to provide a valuable military capability at relatively moderate cost. In an environment where "the *mean value* of our single greatest threat is considerably reduced," but "the *variance* of the collective threat that we are dealing with, planning for, and must counter today is way up," [114] the aggressive but cost-effective development approach used for the F-117A may be just as appropriate now as it has ever been.

Appendix A

EARLY EFFORTS TO DEVELOP LOW RCS VEHICLES

FIRST ATTEMPTS: PARASITE TREATMENTS

HORTEN IX FLYING WING (1944–1945)

This German aircraft was designed to have a sandwich skin with a material derived from charcoal to be used in the core to absorb radar energy.[175, 229] The prototype flown in 1945 did not actually include that feature, however. Although the shape of the aircraft, as seen in Fig. A1, also appears to indicate RCS reduction, the design is in fact similar to earlier Horten sport aircraft for which RCS was not a consideration. Thus, the flying-wing shape did not arise from RCS considerations, although the designers might have used that as an argument in favor of their aircraft.

TYPE XXI U-BOAT SNORKEL TUBES (1945)

Mats of RAM were affixed to the snorkel tubes of a handful of Kreigsmarine Type XXI U-boats very late in World War II to reduce the chances of detection by airborne search radar. This reduced the RCS by approximately 70%

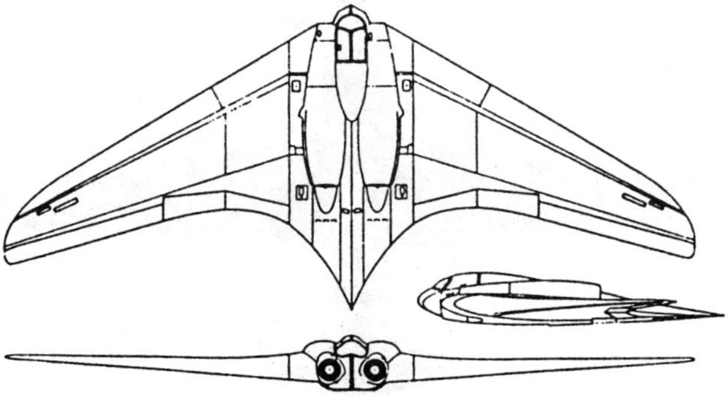

Fig. A1 Horten IX.[189] (Courtesy of Monogram Aviation Publications.)

with a corresponding reduction in detection range of 25%.[26, 189] Considering the limited performance of both the radars of the time, the aircraft carrying the radars, and the added difficulty of detecting the snorkel tubes against the sea surface (rather than in free space), the treatment might have yielded some measurable benefit to the submarines if it had been introduced earlier.

Lockheed U-2 RCS Reduction Attempts (1958–1960)

RCS was not a factor in the design of the U-2. It was designed to achieve survivability simply by flying extremely high. It began overflights of the Soviet Union in 1956 but was quickly discovered by Soviet radars. The United States correctly believed that within a matter of time the Soviets would develop a weapon that could reach it.[186, 197] As a result, various attempts were made to make the U-2 harder to detect. Radar absorbent sheets were affixed to portions of the surface. In another experiment, almost the entire aircraft was coated with printed circuits designed specifically to suppress S-band (2–4 GHz) radar returns.

One of the most exotic treatments, intended to work against relatively low-frequency 70 MHz surveillance radars, is illustrated in Fig. A2. Wires were mounted a quarter-wavelength away from the leading and trailing edges of the wing and tail surfaces, so that the return from the wire would ideally be canceled out by the return from the edge, being 180 deg out of phase with each other.[33, 105]

"The aircraft so equipped was flight tested with negligible results in reduction of radar cross section together with the expected adverse aerodynamic effects...resulting in a considerable loss of cruising altitude," observed Kelly Johnson, speaking at the 1975 Radar Camouflage Symposium at Wright–Patterson Air Force Base.[105] None of the other treatments were any more suc-

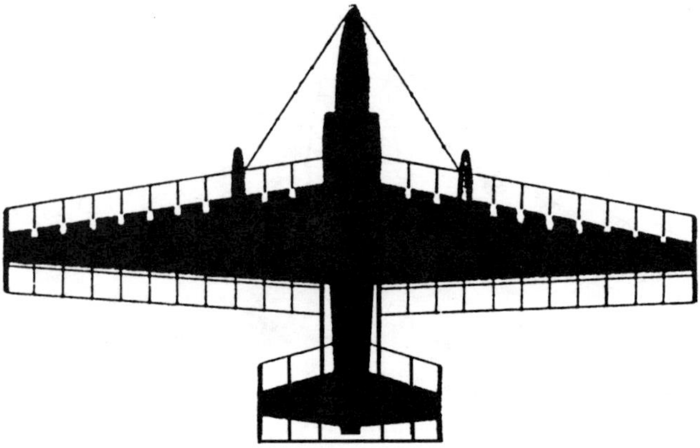

Fig. A2 U-2 with 70 MHz anti-radar "trapeze."[105]

cessful. The treatments designed to work against specific frequencies had little or no effect against other frequencies. Furthermore, major scattering sources such as the engine inlets were unaffected by any of the treatments. At the same time, the weight and drag penalties deprived the U-2 of its vital high-altitude capability, producing a net *decrease* in survivability.[33, 105] These experiences "merely reinforced the writer's [Johnson's] feelings that once an aircraft configuration is set, very little can be done by 'add-ons' to obtain important reductions of the radar cross sections at any important radar frequency."[105]

Teledyne Ryan BQM-34A (Approximately 1960 Retrofit)

A variant of the very successful and versatile Firebee drone (shown in Fig. A3) was flown with a wire mesh screen over the inlet and some RAM applied to various portions of the airframe. RCS was reduced but not to a very low level. According to Teledyne Ryan, "the study showed the limitations of retrofitting an already designed aircraft to reduce its RCS."[256]

North American Hound Dog Air-to-Surface Weapon (1962 Retrofit)

The Hound Dog missile fleet, which entered service in 1960, underwent a fleet-wide retrofit in 1962. The original structure in certain portions of the inlet spike and duct was replaced with radar absorbent structure.[26] The basic weapon (Fig. A4) had few sources of significant scattering in the frontal sector besides the inlet: no canopy and no radar (the guidance system was inertial) and all flying surfaces had highly swept leading edges for supersonic flight.[104] For this reason, treatment of the inlet had a noticeable effect on the weapon's total RCS in the frontal sector and was projected to provide a significant improvement in survivability. A high reliability was also attained. No downtime in more than 10 years of service was attributed to the inlet treatment.[26] However, the inlet geometry of the Hound Dog missile was extremely simple and accessible compared to typical military aircraft. Also, of course, the life of a missile does not consist of many hours of actual flight, but only

Fig. A3 Teledyne Ryan BQM-34A.[256]

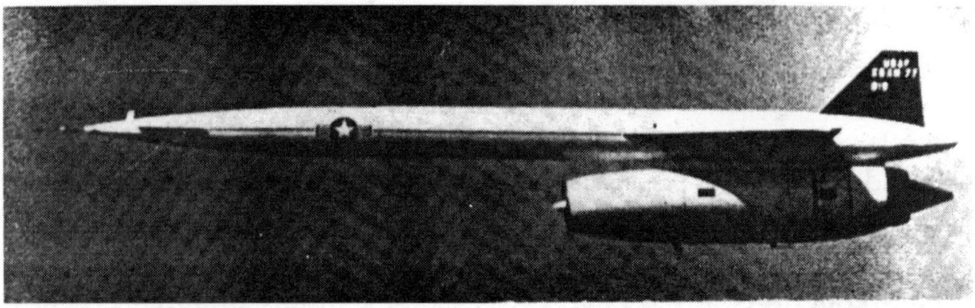

Fig. A4 North American AGM-28 Hound Dog.[104] (Courtesy of Jane's Information Group.)

of being *ready* to fly. This was one of the more successful retrofit efforts, although the effectiveness of it was never tested in combat. The treatment did not affect side or aft-aspect signatures, as these were not considered critical for the Hound Dog's mission.

DESIGNING FOR LOW RCS

LOCKHEED GUSTO 2 (1956–1957 DESIGN STUDY)

Gusto was the name used for a series of studies exploring a possible subsonic, high-altitude, low-RCS replacement for the U-2 after it was discovered that Soviet radars were easily tracking the U-2 during its overflights. Gusto 2, shown in Fig. A5, was claimed to have a very low RCS. However, because of its rounded wing leading edge and only moderate sweep—both features necessary for efficient subsonic flight —there would certainly have been a considerable signature spike not very far from the nose-on aspect associated with the specular return from the wing leading edge. This might have created an appreciable chance of detection, considering that Gusto's intended mission involved long-duration, deep penetrations of Soviet airspace, where hostile radars might not just be ahead of the aircraft but all around it.

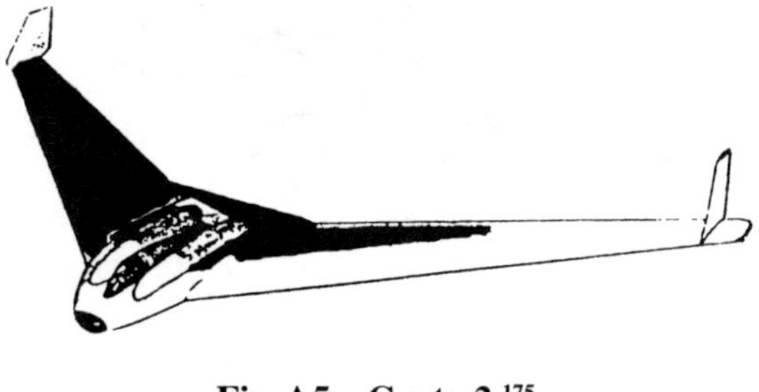

Fig. A5 Gusto 2.[175]

Additional, related studies were conducted by the Lockheed Skunk Works, in conjunction with the CIA, on the effects of speed, altitude, and RCS on survivability. Given the levels of signature reduction that were considered achievable, high supersonic speed in combination with high altitude and some amount of RCS reduction were found to provide the best chances of survival. Therefore, attention shifted away from the long-endurance, subsonic design toward what would become the Lockheed A-12.[175]

LOCKHEED A-12 (FIRST FLIGHT 1962)

The A-12 (Fig. A6) was the CIA-funded predecessor to the better-known SR-71 Blackbird. This Mach 3, high-altitude reconnaissance aircraft was effectively shaped to eliminate specular reflections from both the nose and side aspects. Fuselage shaping was recognized as an important contributor to the side aspect signature, and the design shows considerable emphasis on avoiding vertical or near-vertical surfaces. There were, by this time, standard rules indicating how much a surface should be slanted to reduce its return at a particular aspect.[34] The high-design Mach number allowed planform edges to be sharp and highly swept, both of which characteristics facilitated reduced signatures from the nose-on aspect. The aircraft's construction included special radar absorbent edge structures; radar absorbent paints were also used.[105, 175] The A-12 had a low nose-on signature and extremely low side signature, relative to conventional aircraft.

However, the inlets were one obvious source of significant scattering in the frontal sector. Aside from the issues of internal multiple reflections and compressor face returns, the inlet lips of the A-12 provide excellent examples of *caustics,* where not just one point on the curve but the whole curve may

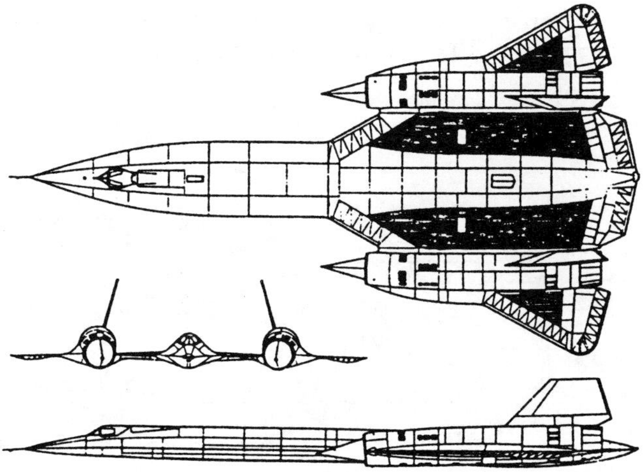

Fig. A6 Lockheed A-12.[175]

diffract toward the receiver (Fig. B8 in Appendix B). Because the inlets are canted slightly down and inward, these caustics would appear slightly to either side of the nose-on aspect and slightly below the horizontal plane. At aspects slightly off the caustics, there would still be two flare spots on each inlet lip (at the farthest and nearest points, respectively). Kelly Johnson, during the development of the A-12, wrote that: "Inlets are the problem in the forward aspect and the exhaust in the rear, as expected." [175] With the demands to withstand high temperatures and to operate efficiently over an extreme speed range, little could be done to reduce the radar returns of the inlets. Shaping the inlets was not considered feasible; the translating-spike inlets posed a tremendous technical challenge, and an axisymmetric shape was considered to be the only workable configuration.[34] Grids or screens were also out of the question, and titanium was the only material from which the inlet lip itself could be made.[105] However, the penalty in forward sector RCS was apparently acceptable for an aircraft that traveled at Mach 3 and at an altitude of around 17 miles. According to Kelly Johnson, "It was not possible to create an aircraft invisible to early warning or missile guidance radars, but in combination with high speed, high altitude, maneuvering, and electronic countermeasures, a very high degree of survivability has been obtained in service." [105]

LOCKHEED D-21 (FIRST FLIGHT 1966)

The D-21 (Fig. A7) was a high-altitude, Mach 3.8 reconnaissance drone intended to be launched from the back of a specially modified A-12 "mother ship." The approach to RCS reduction for the D-21 was similar to that used on the A-12, and it may be assumed that similar results were achieved. The nose-on signature of the D-21, like that of its larger cousin, would have probably been dominated by the inlet. Although it met its speed and RCS design goals, the D-21's career was plagued by various difficulties, and it was canceled in July 1971 after only four operational missions.[175]

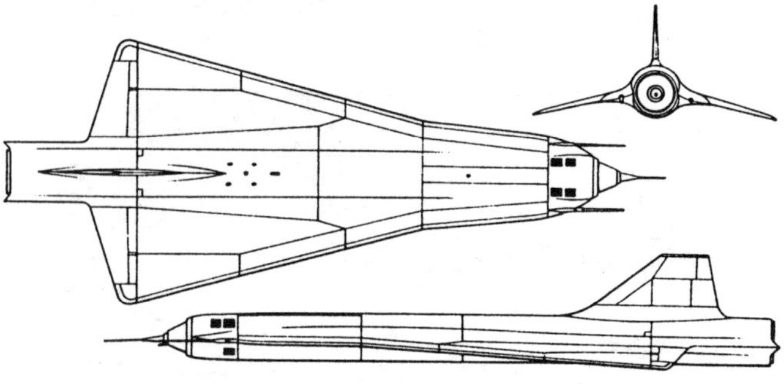

Fig. A7 Lockheed D-21 Drone.[175] (Courtesy of Jane's Information Group.)

TELEDYNE RYAN AQM-91A COMPASS ARROW (FIRST FLIGHT 1969)

Development of this high-altitude reconnaissance drone, illustrated in Fig. A8, was initiated in 1966. Although fairly conventional in its design, the AQM-91A incorporated several RCS-reducing features. Because its mission involved high-altitude flight, special attention was paid to reducing the RCS of the aircraft as seen from below. The vertical tails and fuselage sides were canted inward to eliminate specular reflections from the side aspect at or below the horizontal plane. The engine inlet was lined with RAM and was located on top of the fuselage so that it would be entirely or partially concealed to viewing angles below the horizontal plane. The exhaust nozzle was cooled to reduce infrared signature, and like the inlet, was placed to be hidden by other airframe features at many aspects of observation. The underside of the vehicle was flat and nearly featureless. RAM was applied to the wing leading edges and to some portions of the fuselage.[256] This drone achieved moderately low signatures and was considered successful enough that it provided some impetus for further efforts in signature reduction.[26] However, its real value was the experience that it provided to the designers at Teledyne Ryan. They learned that a conventional, wing–body–tail type of configuration would never be able to achieve really low signatures.[256] Subsequent Teledyne Ryan designs would be much more aggressive.

BOEING AGM-69A SHORT RANGE ATTACK MISSILE (SRAM) (FIRST FLIGHT 1969)

The Boeing short range attack missile (SRAM), shown in Fig. A9, was successfully designed to satisfy a requirement for a very low RCS value in the frontal sector.[26] Relatively little special design or treatment was necessary to achieve this result because of the absence of any major frontal scattering sources in the basic vehicle: no wings, no canopy, no radar (SRAM, like the Hound Dog, is inertially guided), and no engine inlet (SRAM is rocket powered). Treatment consisted of small amounts of lossy magnetic material

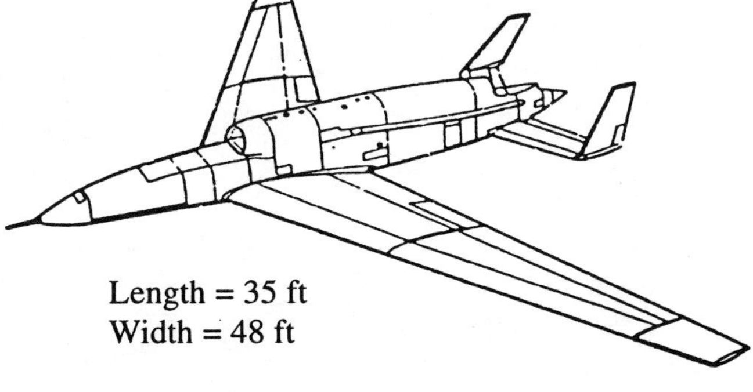

Fig. A8 Teledyne Ryan AQM-91A Compass Arrow.[256]

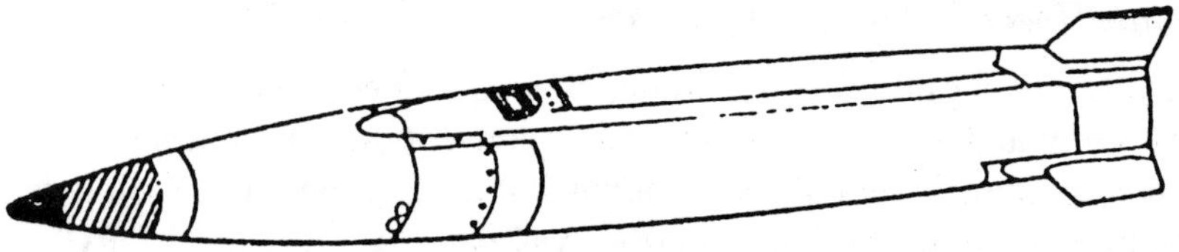

Fig. A9 Boeing AGM-69A SRAM.

in selected areas of the missile. It is interesting that, in 1976, published sources noted that the SRAM "exceeded specification requirements in terms of range, accuracy, radar cross-section and reliability.... Its radar 'signature' is said to be no larger than that of a machine-gun bullet." [104]

McDonnell Quiet Attack Aircraft (1972–1973 Design Study)

The Quiet Attack aircraft (Fig. A10) was designed by McDonnell Douglas under contract to the Office of Naval Research for minimum radar, infrared, and acoustic signatures. In keeping with the concept of a slow, quiet, covert attack aircraft, it embodied a much more extensive effort to control engine signatures than was typical for earlier low RCS aircraft. This effort included hidden inlets; a high-bypass, tip-driven turbofan engine; and a special plug nozzle.[257] Characteristics of the design are summarized in Table A1. In weight,

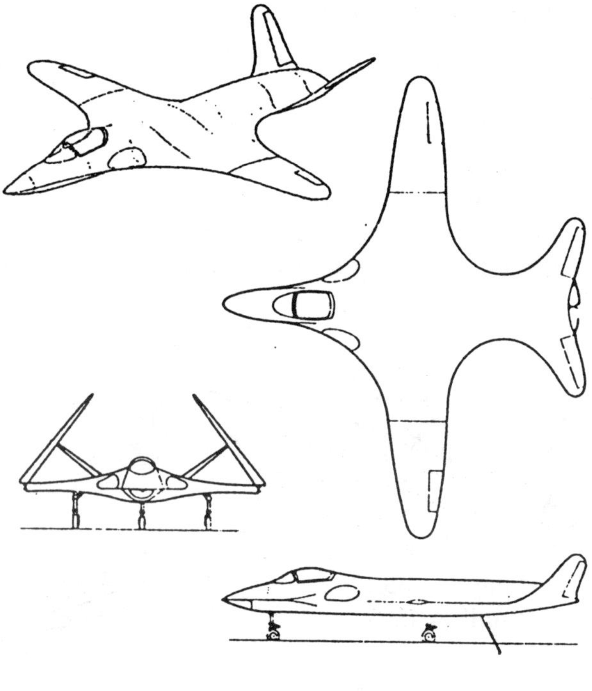

Fig. A10 Quiet Attack aircraft.[257]

the aircraft is very close to Have Blue. Boeing also participated in the Office of Naval Research study, although information on the Boeing concept could not be located when researching for this book.[119]

The Quiet Attack aircraft's shape was carefully blended so as to slant all surfaces quite far from the vertical, indicating the typical appreciation for the importance of specular reflection. The shaping approach is summarized: "The most important technique for RCS reduction is a blended shape. This means no straight leading or trailing edges, no flat sides on the fuselage, no single vertical tail...."[257]

However, this approach leads to problems with edge returns. From the top view in Fig. A10 the continuously curved planform edge apparently would have created a "flare spot" (a part of the edge perpendicular to the line-of-sight and therefore producing a relatively strong diffracted return). The flare spot would have occurred for nearly every viewing aspect.

Teledyne Ryan Low RCS Vehicle (1973–1974 Design Study)

The Remotely Piloted Vehicle Special Projects Office at Wright–Patterson Air Force Base, Ohio, conducted studies and tests in conjunction with Teledyne Ryan and other companies of low RCS drones. Having learned from experience with the AQM-91A Compass Arrow that a conventional, wing–body–tail arrangement would not achieve very low signatures, Teledyne Ryan advanced to a more aggressive approach involving simpler planforms with fewer surfaces and edges. The Teledyne Ryan Low RCS Vehicle, illustrated in Fig. A11, was to consist of a small metal centerbody surrounded by a large amount of lossy dielectric material. The configuration was all-wing with a simple delta planform and inward-canted twin vertical tails. The engine and equipment were buried in the fairly thick wing root. Tail surfaces and other features were to be made of "radar-transparent" materials. Overall dimensions would be approximately 18 ft in overall length and 21 ft in wingspan.[133, 256]

Table A1 McDonnell Quiet Attack aircraft characteristics[257]

Length	44.6 ft
Wingspan	56.5 ft
Take-off weight	14,410 lb
Empty weight	9360 lb
Fuel weight	3415 lb
Wing area	400 sq ft
Aspect ratio	8
Take-off thrust to weight ratio	0.44
"Quiet speed" at 2500 ft altitude	112 kn
Maximum speed at 2500 ft altitude	400 kn

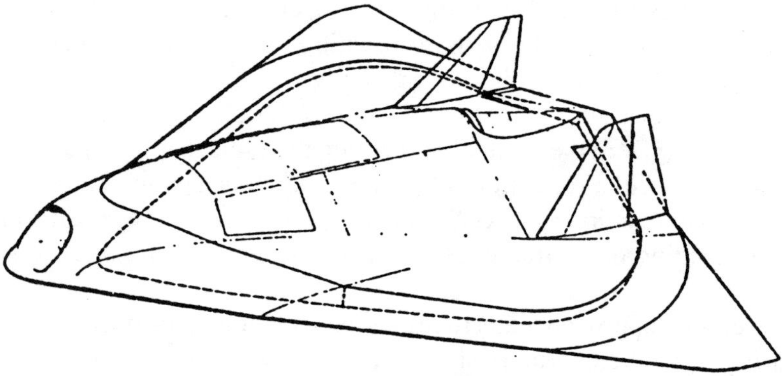

Fig. A11 Teledyne Ryan Low RCS Vehicle.[256]

The vehicle was designed to a very demanding set of RCS goals, which were derived from extensive threat analyses and covered a broad range of frequencies corresponding to the various elements of the latest Soviet air defense systems. (Air defense systems are discussed in more detail in reference.[189] Model tests showed impressive RCS results, but the approach was considered impractical, and the vehicle still did not meet its RCS goals at all frequencies of interest.[133, 235, 256] Practical shortcomings included a very small useful volume relative to the total size of the vehicle and difficulties in making certain components truly "radar-transparent."

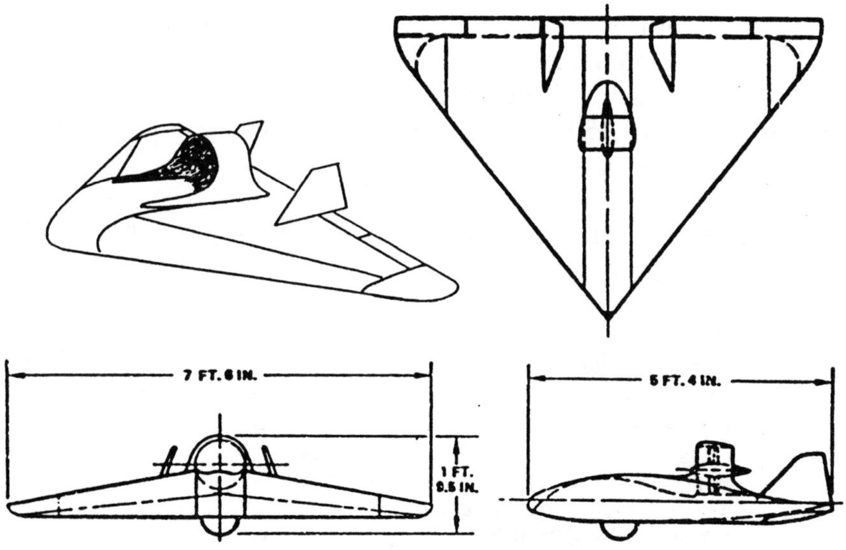

Fig. A12 Teledyne Ryan Mini-RPV.[256]

TELEDYNE RYAN MINI-RPV (1974–1975)

The 7.5-ft wingspan Teledyne Ryan Mini-RPV followed the same shaping approach as the Low RCS Vehicle, but it specifically avoided reliance on RAM. Several versions were proposed with variations in the tradeoff between radar treatment, countermeasures, and overall system cost. As shown in Fig. A12, it was powered by a ducted propeller on the top side of the vehicle, and in some variants it was intended to use wire screens over both ends of the duct, as well as over the sensor package on the vehicle's underside. In RCS testing, Mini-RPV models met or exceeded signature goals that were based on achieving survivability through a combination of RCS reduction *and countermeasures*. Some Mini-RPVs were built and flown, although it is unknown whether any of these represented a low-RCS configuration or whether any in-flight RCS testing was ever performed.[104, 133, 235, 256]

Neither the Mini-RPV nor the Low RCS Vehicle achieved sufficiently low signatures to avoid detection completely while performing a useful military mission. However, they did provide what was perhaps the first credible indication that such a goal could be accomplished. Many radar systems did not perform at all well against these targets.[235] The lesson was not wasted. Several of the people involved in these projects at the Air Force's RPV office later played key roles in formulating and managing the Have Blue program.[188, 234]

Appendix B

Development of RCS Prediction Methods

Introduction

The importance of RCS reduction has been recognized since before World War II. Most early attempts at RCS reduction consisted of little more than applying radar absorbent materials to all or part of a vehicle. These efforts generally met with little or no success. By the late 1950s, Air Force officials and contractors realized that a very low RCS would not be achieved through "parasite treatments" alone (paints, coatings, and other additions to existing aircraft) but would have to be "designed in" from the beginning. Designers needed a way to evaluate the effect of various features on the RCS of the vehicle during the design process. This was accomplished by analysis, testing of subscale models, or ideally through a combination of the two.

Having, in general, little knowledge of electromagnetic theory, aircraft designers first turned to experimental techniques. Aircraft RCS measurements had been made as early as the end of World War II, and the first dedicated RCS measurement ranges were built in the 1950s. RCS model testing was used in the design of certain aircraft, including the Lockheed A-12 (the CIA-funded predecessor to the better-known SR-71). However, large RCS improvements were still hard to achieve. Although designers could test and modify through trial and error, theoretical knowledge or experience to guide their attempts at RCS reduction was not available. Analytical prediction techniques capable of providing practical insight for realistic aircraft shapes were needed.

The fundamental laws that govern the propagation and scattering of electromagnetic waves were formulated by James Maxwell more than 100 years ago. However, exact mathematical solutions to Maxwell's equations are very few and are limited to simple shapes such as a sphere, ellipsoid, infinite wedge, and a few others. Thus, to predict the radar return from an object of arbitrary shape such as an aircraft, designers had to resort to approximate techniques that captured the dominant scattering mechanisms under certain sets of simplifying assumptions. Considerable effort has been devoted to developing such techniques.

Definition of Radar Cross Section

When an object is illuminated by radar waves, energy is scattered in all directions with various degrees of intensity. In most cases of practical interest, referred to as monostatic, the radar receiver is in the same location as the transmitter; producing an amount of radiation scattered back toward the source. This is called backscatter. The present discussion is confined to this case.

Radar cross section is defined as the ratio of the scattered power density in a given direction (usually the backscatter) to the incident power density normalized so as to be independent of the distance R at which the scattered power is measured:

$$\sigma = \lim_{R \to \infty} 4\pi R^2 |S_s|/|S_i| \tag{1}$$

where S_s and S_i denote the energy flux of the scattered and incident radiation respectively. Because the scattered field is radiated spherically outward from the object, its intensity decays as $1/R^2$, so the power density ratio is multiplied by R^2 to eliminate the dependence on R, making σ a property of the object and not the measurement distance. The RCS does depend on the frequency of the incident radiation, on the aspect or orientation of the object relative to the incident radiation, and on the polarization of the incident wave. (In other words, the orientation of the respective magnetic and electric field components of the incident wave; see Fig. B1.)

Units of measurement for RCS are square meters. However, because objects of interest (ranging from birds to significant geographical features) have radar cross sections that span a range of several orders of magnitude, measurement in decibels is the most convenient. In this form, the common unit of measurement is dBsm, or decibels relative to a one square meter reference area:

$$\sigma_{\mathrm{dBsm}} = 10 \cdot \log_{10}(\sigma) \tag{2}$$

A change of 10 dB corresponds to one order of magnitude. The maximum range at which an aircraft can be detected by radar is given by the well-known radar range equation

$$R_{\max} = \left[P_t G^2 \lambda^2 \sigma / (4\pi)^3 P_{\min}\right]^{\frac{1}{4}} \tag{3}$$

where P_t is the transmitted power, G is the radar antenna gain, λ is the wavelength, and $P_{\min}$ is the minimum detectable returned signal power. This equation shows that detection range is proportional to the fourth root of the radar cross section. Therefore a very large change in RCS is needed to achieve a tactically significant reduction in detection range.[123, 213]

ELECTROMAGNETIC SCATTERING

Maxwell's equations in differential form are given here[134]:

$$\nabla \cdot \boldsymbol{E} = \rho/\varepsilon_0 \tag{4}$$

$$\nabla \cdot \boldsymbol{B} = 0 \tag{5}$$

$$\nabla \times \boldsymbol{E} = -\,\mathrm{d}\boldsymbol{B}/\mathrm{d}t \tag{6}$$

$$\nabla \times \boldsymbol{B} = \mu_0(\boldsymbol{J} + \varepsilon_0\,\mathrm{d}\boldsymbol{E}/\mathrm{d}t) \tag{7}$$

where $\boldsymbol{E}$ is the electric field intensity, $\boldsymbol{B}$ is the magnetic field intensity, ρ and $\boldsymbol{J}$ are the charge and current densities respectively, and ε_0 and μ_0 are respectively the permittivity and permeability of a vacuum. The equations can be readily manipulated to show that electromagnetic waves exist, and that in a vacuum they propagate at a speed where

$$c = 1/\sqrt{\varepsilon_0\mu_0} \cong 3 \times 10^8 \text{ m/s} \tag{8}$$

The waves consist of an electric field component $\boldsymbol{E}$ and a magnetic field component $\boldsymbol{H}$ that are normal to each other and normal to the direction of propagation. Furthermore, the two fields are periodic and in phase with each other, as illustrated in Fig. B1.

Note that, in the notation, the magnetic induction $\boldsymbol{B}$ has been replaced by the magnetic field intensity $\boldsymbol{H}$. In free space these are quite simply related to each other by the relationship

$$\boldsymbol{B} = \mu_0\boldsymbol{H} \tag{9}$$

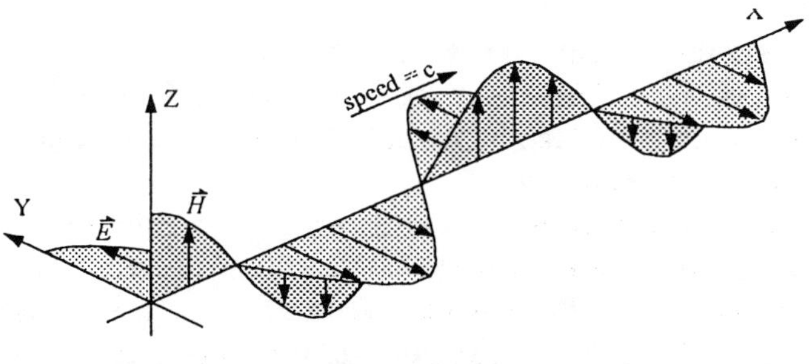

Fig. B1 Electromagnetic waves.

The total energy flux of the wave is given by the cross product of the electric and magnetic components, also known as the "Poynting vector": [134]

$$S = E \times H \tag{10}$$

The Poynting vector is always oriented in the direction of propagation of the wave.[134]

If an electromagnetic wave encounters an object, then electrical currents are induced on the object's surface. These currents in turn produce the scattered wave. The interrelationship between the incident wave, the induced currents, and the scattered wave is defined by Maxwell's equations together with the boundary conditions on the target's surface. However, as noted previously, exact solutions to this coupled system of equations do not exist for practical aircraft shapes, so designers must resort to assumptions that simplify the problem.

Simplifying Assumptions

Designers often assume, as a starting point, that the entire surface of the object in question is a "good conductor," in other words metallic, with no gaps or discontinuities. This assumption leads to a relatively simple boundary condition. The effects of radar absorbent materials or disruptions in the conductive surface are considered separately. The next assumption regards the wavelength of the incident radiation relative to the physical size of the object and its important features. Three major scattering regimes exist: low-frequency (or Rayleigh scattering) in which the wavelength is assumed to be much larger than the object; resonant scattering, in which wavelength is comparable to the size of the object (or important physical features of the object); and high-frequency scattering, in which the wavelength is assumed to be much smaller than any feature of the object. Furthermore, within each scattering regime there are various theories that capture different sets of scattering mechanisms. Which theory is chosen depends on which set of mechanisms is expected to be dominant in a particular situation.

Low-Frequency Methods

If the incident radiation has a wavelength much larger than all characteristic dimensions of the object whose RCS is being calculated, then the phase and amplitude variation of the incident field over the length of the object is negligible, as illustrated in Fig. B2.

The field therefore appears as a spatially uniform field varying only in time. The variation is slow enough so that the charge distribution can be found for some reference value of the incident field strength, as if the incident field were constant in time. This assumption is referred to as a "quasi-steady" assumption and allows the techniques of electrostatics to be used. Mutual

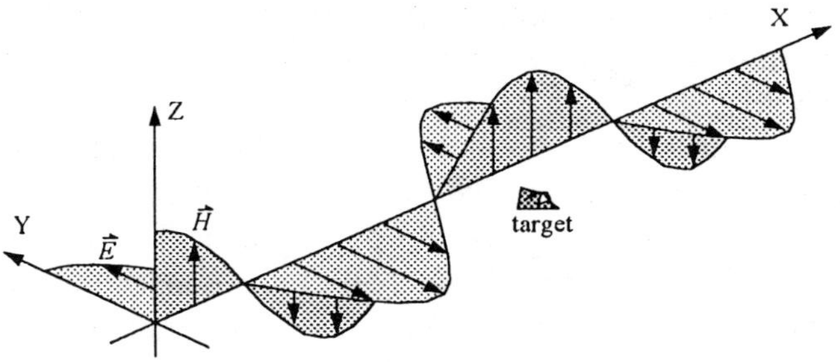

Fig. B2 Low-frequency case.

influence (which is the effect of the charge at one point of the body, on the field at another point) is important, but electrodynamic effects are not, and the problem is relatively tractable. Once the charge distribution is found for some reference value of the incident electric field, its "shape" does not change, but the magnitude of the entire distribution oscillates in response to the modulation of the incident field. The dipole and higher order multipole moments of the reference charge distribution can be calculated, and the radiation patterns associated with these oscillating multipoles are well known, so the scattered field is readily determined.[123, 134]

Unfortunately this fairly simple case does not apply to most radar problems of practical interest. The assumptions of low-frequency scattering may be considered fairly accurate if the phase variation of the incident wave over the length of the object is, perhaps, $\pi/8$ radians or less.[123] This corresponds to a maximum size of 1/16 the wavelength. The longest-wave radars are VHF early warning radars, which have a wavelength between 1–3 m (Ref. 239). So it can be seen that low frequency methods can only be applied to objects with characteristic dimensions of a fraction of a meter, which clearly does not include manned aircraft.

Medium- (Resonant) Frequency Methods

When the characteristic dimensions of an object are of the same order as the incident wavelength, designers must account for the phase variation of the incident field over the length of the object, and for electrodynamic effects (such as the magnetic fields induced by currents on the body, as well as the electric fields induced by the charges). In the low-frequency case, designers must also account for mutual influence. The charges and currents at every point on the object affect the fields at every other point on the object. The solution procedure is very similar to aerodynamic panel methods. A matrix of mutual influence coefficients is set up between a selected number of points on the surface of the body. The current distribution is expressed in terms of a series. Enforcing the boundary conditions on the surface of the object yields a system of linear equations from which the series coefficients that define the current distribution can be determined. This method is called the Method of Moments (MOM).

For accuracy, control points should be spaced at intervals of around one-fifth wavelength or less. As of 1985, to apply the method to objects up to roughly 10 wavelengths in size was considered practical.[123] Advances in computational power have undoubtedly increased this upper limit in the last 10 years. However, apart from the computational issue, applying MOM to very large objects—or very small wavelengths— is limited because MOM does not capture some of the physical phenomena of interest for the high frequency regime. In MOM the influence of each point on every other point is taken into account, but only in reference to the spatial positioning of the two points relative to each other. For example, the field at point A is induced by a given current density at point B.

At very high frequencies, that kind of interaction becomes largely negligible. However, other kinds of interaction become important, which depend not only on the relative position of two points but *on the nature of the surface of the body between the two points.* Close points may have very little influence on each other, while seemingly very widely separated points—for example at opposite ends of the body—may interact strongly through phenomena such as surface traveling waves or creeping waves.

Nevertheless, MOM is very useful for analyzing portions of aircraft such as edges or inlet ducts, which have characteristic dimensions of only a few wavelengths or less. In the early 1970s, Teledyne Ryan used MOM to optimize the edge treatments on its Low RCS Vehicle design.[133] This is further discussed in Appendix A. Application of this method was possible because the edges were treated as a two-dimensional problem, greatly reducing computational requirements relative to a full three-dimensional problem. Since the mid-1980s, computational capabilities have made analysis practical for more general three-dimensional features using MOM, then merging the results into the solution for the complete aircraft.[34]

High-Frequency Methods

Most radar frequencies fall into the high-frequency regime when the target is a full sized aircraft. High frequency methods are therefore of particular interest for practical RCS prediction. Many approaches to high-frequency RCS prediction have been used. The most important are in the categories of geometric optics, physical optics, Geometric Theory of Diffraction, Physical Theory of Diffraction, and more recent developments that have improved upon the two latter theories. Specifically, these are the Method of Equivalent Currents and the Incremental Length Diffraction Coefficient.

In general, the induced currents on the target surface depend on the total field—the incident field plus the scattered field. Thus, the current distribution and the scattered field must be found together, as each influences the other. However, if all characteristic distances on the target are large compared to the incident wavelength, then contributions to the scattered field from distant features of the target may be neglected. The currents at each point become a function only of the incident field and the local scattered field.

"Geometric" methods use this result in a qualitative way, allowing the total scattering to be thought of as the sum of all identifiable local contributions. "Physical" methods go a step further and use the high-frequency assumption to decouple the surface current distribution from the total scattered field. The surface current distribution is found first. Once this is known, the surface integral that determines the scattered field may be evaluated.

GEOMETRIC OPTICS (GO)

The oldest high-frequency method is geometric optics (GO) and consists simply of tracing the paths of rays, thus bypassing the need to determine surface currents. GO can account for the mechanisms of specular reflection (rays "bouncing" off an object, as in Fig. B3) and of refraction (rays bent at the interfaces of mediums with different optical properties, also shown in Fig. B3). GO has been used for a long time in the design of telescope lenses and other optical systems.

Geometric optics is very good for finding the specular reflection from a body with compound curvature at the specular point. However, when one or both principal radii of curvature is infinite at the specular point, GO predicts an infinite value of the radar cross section, which is incorrect.[123]

An advantage of GO is that it can actually calculate some effects from radar absorbent materials, because it does capture the phenomena of refrac-

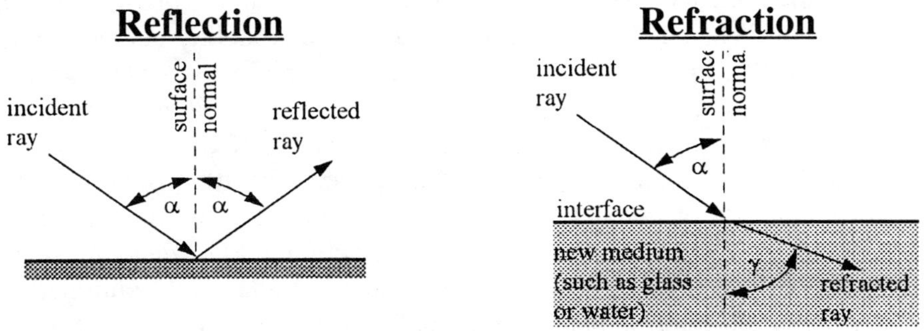

Fig. B3 Reflection and refraction.

tion. Rays can be traced through multiple reflections and/or refractions until they finally are scattered away from the vehicle. This fact also makes GO reasonably well suited to the calculation of backscatter from inlet ducts, where there are multiple reflections from the duct walls and from the engine face. RCS prediction codes developed at Northrop in the late 1960s and early 1970s used GO for this purpose. More recently, as noted, MOM has replaced GO in such applications.

Physical Optics (PO)

The next method, physical optics (PO), actually finds the surface currents on the body and performs a surface integral around the body to find the scattered radiation resulting from the surface currents. However, determination of surface currents is only approximate. Specifically, the current at each point on the body is assumed to be identical to the current in an infinite plane at the same orientation as the point of interest. Thus, any interactions between different parts of the body are neglected. Nevertheless, PO permits reasonably accurate predictions in certain cases where GO fails. Specifically, PO correctly finds the strength of the specular reflections from flat surfaces or those that possess curvature in only one direction. These are cases where GO would predict an infinite cross section.

Because PO determines the scattered field based on integrating the finite currents over the finite surface of the body, PO always yields a finite solution. Furthermore, PO can be used to find the scattering from a surface at nonspecular angles. Results must be treated with care, however, because at angles significantly removed from specular, the scattering may be dominated by mechanisms that are not captured by the PO approximation. For example, because edges form the limits of integration for a surface, edge contributions are present in the PO integral. However, those contributions do not include the effect of "nonuniform" currents that exist near the edges. Such currents are dealt with in the diffraction theories.

An interesting situation in which PO does not accurately predict the scattering occurs when a shadow boundary falls on a smoothly curved surface (Fig. B4). In this case, PO hypothesizes the existence of an edge at the shadow boundary, which is incorrect. GO would be a better method in that case.

This failing may have considerable practical significance as shown in Fig. B5. Any time a smooth airfoil is illuminated from the frontal region, a shadow boundary on one or both of the upper and lower wing surfaces exists. Gupta and Burnside in 1987 developed a correction to PO, which allows the erroneous edge contribution at a smooth shadow boundary to be eliminated.[74]

An important point is that physical optics is not trivial. Although determining surface currents is quite simple, the integration to find the scattered radiation may be quite challenging. Closed-form solutions for the PO scatter-

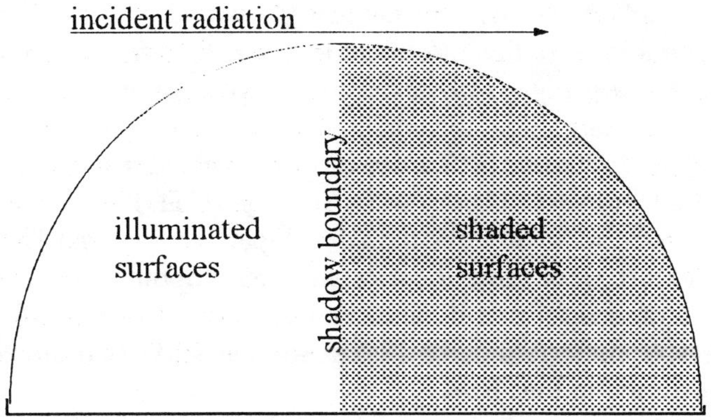

Fig. B4 Smooth object with shadow boundary.

ing from a polygonal flat plate were found independently by several researchers in the mid-1970s. These included W. B. Gordon, who published his results through IEEE in 1975; Kenneth Mitzner at Northrop; and Bill Schroeder at Lockheed.[69, 177, 186, 204]

This was an important development because many of the subsequent methods that capture higher-order phenomena do not provide a total solution but rather a correction to a "basic" solution found by PO. So it is still necessary to find the PO solution by approximating any shape as a collection of flat panels and then using the expressions developed by the aforementioned researchers.

DIFFRACTION

Neither GO nor PO address the mechanism of diffraction from an edge. If an object is shaped to avoid specular reflections from surfaces, then edge diffraction may be the next most significant scattering mechanism. Diffraction occurs anywhere that the surface slope changes rapidly enough to cause

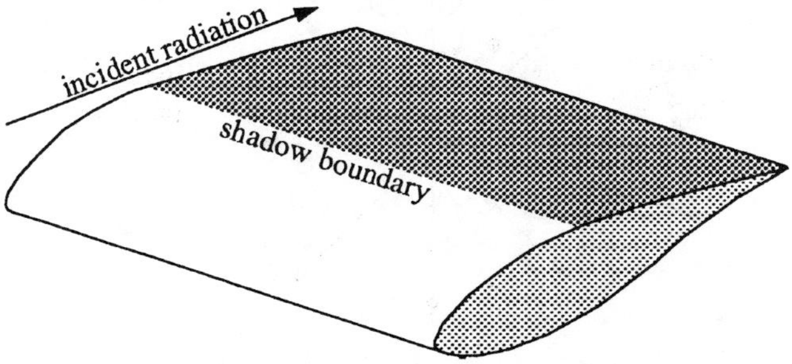

Fig. B5 Smooth airfoil illuminated nose-on.

a perturbation relative to the currents that PO would predict. Generally, this happens when a local radius of curvature is small compared to the incident wavelength. A sharp edge is the extreme example. If the surface currents are interrupted by a sharp edge, the current has no place to go except *along* the edge. During the 1950s two diffraction theories evolved in parallel. J. B. Keller developed the Geometric Theory of Diffraction (GTD) in the United States, and P. I. Ufimtsev in the Soviet Union developed the Physical Theory of Diffraction (PTD). Both of these theories were built upon the exact solution of Maxwell's equations for two-dimensional scattering from an infinite wedge, found by J. A. Sommerfeld in the 1890s. Sommerfeld's two-dimensional solution is illustrated in Fig. B6.

Sommerfeld determined the intensity of the scattered radiation as a function of the scattering direction ϕ_s, the incident direction ϕ_i, and the wedge half-angle α.

Geometric Theory Of Diffraction (GTD)

J. B. Keller extended Sommerfeld's results to three-dimensional and arbitrary angles of incidence by stating that the *normal component* of the scattered radiation behaves according to Sommerfeld's solution, while the tangential component along the edge is unchanged. This provided a solution on a specific locus of scattering directions consisting of a cone whose half-angle is the same as the angle between the incident ray and the edge, shown in Fig. B7.

Keller devised a series of diffraction coefficients that model the three-dimensional extension of Sommerfeld's exact solution and form a convenient way to account for edge diffraction.[117, 123]

Like GO, GTD fails in certain cases where an infinite number of the scattered rays are parallel. In the case of edge diffractions, these directions are called *caustics*. Caustics occur when all the points along some section of

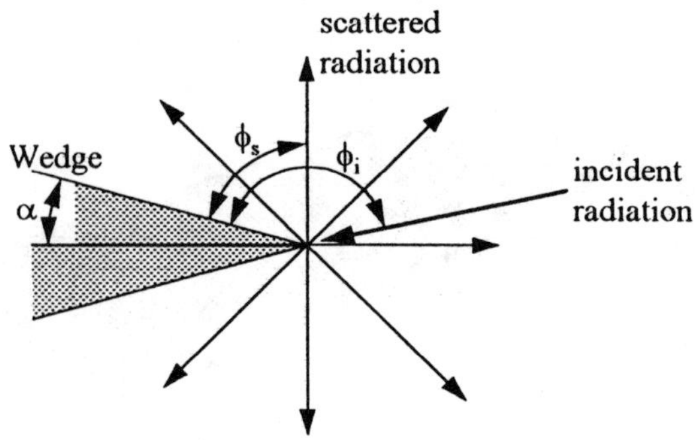

Fig. B6 Two-dimensional diffraction from a wedge.

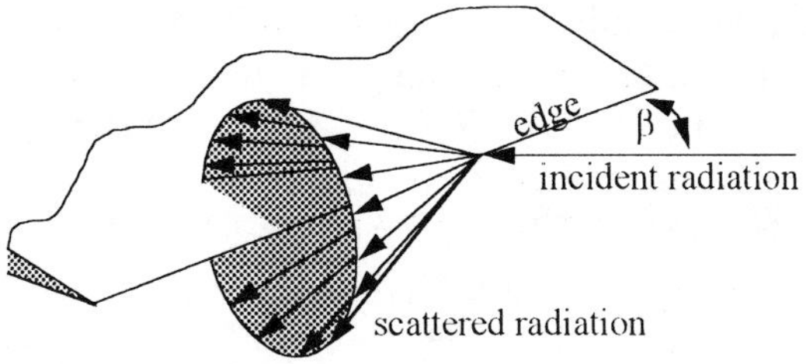

3-Dimensional Diffraction - strongest scattering is on a cone whose half-angle is equal to β, the angle formed by the incident radiation and the edge. This cone is referred to as the "Keller Cone."

Fig. B7 Three-dimensional diffraction by an edge.

nonzero length along a curved edge diffract toward the receiver with the same phase. For backscatter, this requires that the section of curved edge be perpendicular to the incident radiation. Caustics may be encountered, for example, when calculating the nose-on radar signature of an aircraft with a sharp axisymmetric inlet lip, as in Fig. B8.

In these cases, the entire ring of the inlet lip is illuminated, rather than just a single "flare spot," which is often encountered on curved edges. While GTD does not correctly predict the value of the RCS on caustics, it is probably large and caustics should therefore be avoided on low RCS vehicles.

Physical Theory of Diffraction (PTD)

Ufimtsev devised coefficients analogous to Keller's, which do not model the exact solution, but rather the *difference* between the exact solution and the solution found by Physical Optics. Thus, designers would first find the

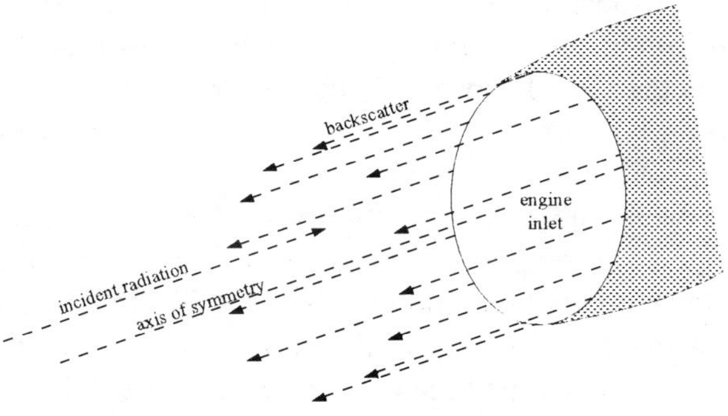

Fig. B8 Caustic associated with axisymmetric inlet lip.

scattering using PO, then apply Ufimtsev's method as a correction. PTD has some advantages relative to GTD, for example, it yields a solution on caustics.[237]

Ufimtsev published the *Method of Edge Waves in the Physical Theory of Diffraction,* a collection of several of his important results, as an unclassified Soviet technical paper in 1962. The U.S. Air Force Systems Command's Foreign Technology Division translated *Method of Edge Waves...* in September 1971 as part of its technical intelligence exploitation program and made it available to U.S. military aircraft companies. However, Ufimtsev's work was already known in some research centers as evidenced by references contained in various Northrop radar cross section reports during the late 1960s and early 1970s.[39, 132, 176, 185] Lockheed made use of Ufimtsev's diffraction theory in late 1975 or 1976 to improve the accuracy of its ECHO RCS prediction program. The initial version of ECHO only calculated the physical optics solution.[34, 186]

Neither PTD, as presented in *Method of Edge Waves...*, nor GTD is valid except on the Keller Cone. In the context of practical RCS prediction, this means that the methods would only calculate the scattering from those edges, if any, whose diffraction cones contained the observation point. The contributions from all other edges would be neglected.[123, 176, 237]

However, the strongest diffraction is on the Keller Cone, so these theories do give the designer an important shaping principal: avoid having sharp edges whose Keller Cones are directed toward a likely detector. For the monostatic case, when the radar transmitter and receiver are collocated, edges that are normal to the direction of incident radiation should be avoided. If there is a curved edge, some portion of it will quite likely be perpendicular to the incident radiation and will therefore diffract toward the receiver; referred to as flare spots.

These considerations are the basis for edge management, an important practical principle in vehicle RCS reduction. To avoid detectable signatures except in a small number of carefully chosen aspects, designers must avoid curved edges and orient the straight edges perpendicular to noncritical aspects. The nose-on aspect is often considered the most critical. To minimize signatures in the frontal sector, all edges must be swept as much as possible.

The diffraction theories discussed do not, however, provide any quantitative information on how much a given edge should be swept or how much RCS reduction is achieved by sweeping an edge a given amount. Both diffraction theories were subsequently improved. The Method of Equivalent Currents, developed and refined by various researchers from the late 1950s through the 1980s, extended the applicability of GTD to observation points that do not lie on the Keller Cone. In the early 1970s, Kenneth Mitzner of Northrop Corporation developed the concept of Incremental Length Diffraction Coefficients (ILDCs) that similarly extended the applicability of PTD.

METHOD OF EQUIVALENT CURRENTS (MEC)

A key feature of both geometric optics and the Geometric Theory of Diffraction is that the radar cross section is determined from a ray description of the scattering: the relevant scattering mechanisms are described in terms of rays, and the divergence of the scattered rays determines the intensity of the radar return at an observation point in the far field. This bypasses the step of finding the currents induced by the incident radiation on the object's surface. Thus, the geometric methods are the simplest to use in cases where the ray-tracing models exist and accurately capture the dominant scattering mechanisms. However, geometric methods do not help when an infinite number of rays are returned in the same direction (specular reflection from cylindrical or flat surfaces and edge diffraction on caustics); or when there is no appropriate ray description of the scattering, (such as for scattering directions that do not lie on the Keller Cone).

The Method of Equivalent Currents addresses these limitations by supposing the existence of a finite current concentrated in hypothetical filaments along each edge of an object, as illustrated in Fig. B9.

The finite currents can be used to calculate the scattered field in any direction and will always yield a finite result. MEC was applied to certain special cases in the 1950s and 1960s, and was more fully developed by Arie Michaeli in the 1980s.[123, 173]

Unfortunately, the magnitude of the edge current depends on the scattering angle and is therefore a fictitious quantity. The actual currents on the object obviously cannot be affected by the location of the observer. The reason for this can be seen by looking at the foundation of MEC, which uses currents in a *filament* to simulate actual edge currents (Fig. B10). Since the cross section of the filament is considered to be a point, it is obvious that the filament is incapable of introducing any dependence of the scattered field on

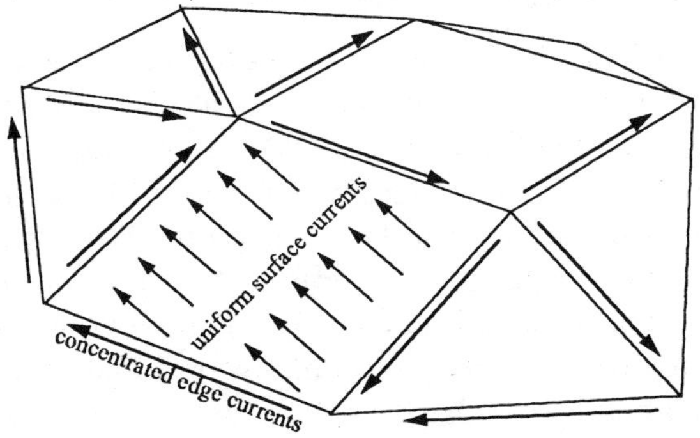

Fig. B9 Method of Equivalent Currents (MEC).

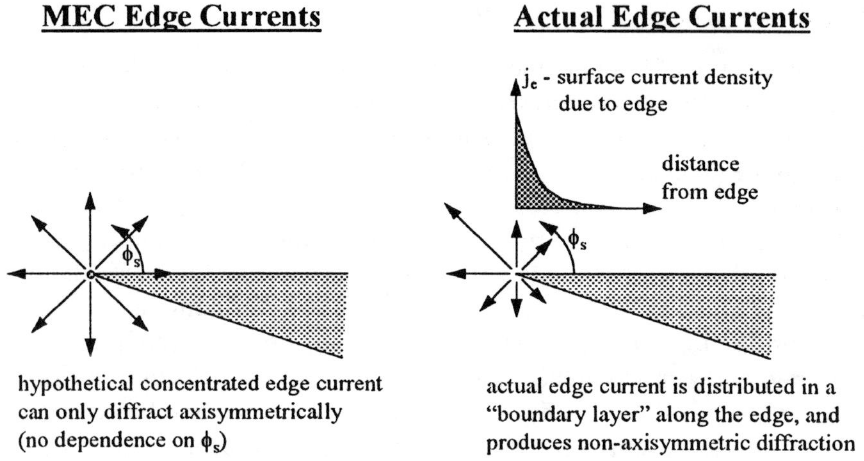

Fig. B10 Difference between MEC (axisymmetric) and actual edge scattering.

ϕ, the azimuth angle relative to the edge of interest. Because the actual diffraction from an edge does have a dependence on the azimuth angle, ϕ, the only way for MEC to capture this dependence is by using a different current in the filament for different values of the scattering azimuth. The current is found by matching the exact solution on the Keller Cone for the scattering azimuth of interest, and the current is then used to find the diffracted field for other values of ß at the same value of ϕ_s. Calculations at different values of ϕ_s require a new value of the current to be determined. Nevertheless, MEC is a very useful extension of GTD. It is possible that improved ways of modeling the edge currents will be (or have been) developed to capture the azimuthal variation of the diffracted radiation.

Incremental Length Diffraction Coefficient (ILDC)

In April 1974, Kenneth Mitzner of Northrop Corporation's Aircraft Division published his *Incremental Length Diffraction Coefficients* (ILDC), which extends the applicability of Ufimtsev's PTD in a manner that is analogous to what MEC does for GTD, allowing the diffracted radiation to be calculated even for scattering directions that do not lie on the Keller Cone.[123, 176] However, because Ufimtsev's diffraction coefficients do take into account the distribution of current near the diffracting edge (at least implicitly), the method of ILDCs does not require different solutions for different scattering angles.

For historical interest, and also because it concisely summarizes important features of Mitzner's method, the following description is quoted directly from his paper:

> In calculating the scattering from three-dimensional bodies with edges, it is frequently meaningful and useful to consider the scattering associated with an incremental length of the edge and to describe this scattering in terms of an Incremental Length Edge Diffraction Coefficient (ILEDC). In this report the theory of the ILEDC is developed, taking into account the actual distribution of surface current near the edge. The theory is illustrated by applying it to the problem of scattering from a perfectly conducting polygonal plate. The Incremental Length Diffraction Coefficient (ILDC), which is the generalization of the ILEDC for linear scattering features other than edges [such as rounded corners], is also treated....
>
> The 2-D DC [two-Dimensional Diffraction Coefficient] provides a simple and compact way of expressing the well-known edge diffraction results of Ufimtsev and Keller.
>
> The ILDC [Incremental Length Diffraction Coefficient], which is defined for all directions of incidence and scattering, is shown to be a generalization of the 2-D DC, which is defined only for certain combinations of incident direction and scattering direction. Wherever the 2-D DC is defined, it is equal to the corresponding ILDC.[176]

Although Mitzner still neglected corner effects and the interactions of multiple edges with each other, his results compared well with experimental data for the polygonal plate of his example. ILDC represented an important advance in the ability to design low RCS aircraft. Like Ufimtsev's PTD, the ILDC solution does have to be added to the PO solution to yield the total scattered field. For this reason, Mitzner published *Studies in Physical Optics*[238] concurrently with and as a complement to his *Incremental Length Diffraction Coefficients. Studies in Physical Optics* presents a variety of solutions and/or approximations that are useful in finding the basic PO solution, including most notably a closed-form solution for the PO scattering from a polygonal flat plate.

Mitzner's work was part of an extended research program conducted by Northrop under sponsorship of the Air Force Avionics Laboratory at Wright–Patterson Air Force Base since the mid-1960s. By 1974 Northrop had developed a computer program called GENSCAT that incorporated all of the theories available at that time—GO, PO, GTD, PTD, and ILDC—to calculate the RCS of complete aircraft configurations. GENSCAT had special routines to calculate the effects of radar absorbent coatings, and to trace the multiple reflections into and out of inlet ducts. GENSCAT demonstrated reasonably good agreement with test data in predicting the RCS of an F-4C aircraft. However, inputting the geometry of the F-4C into the computer program was a significant project in itself.[132]

LIMITATIONS

Each of the theories just described has limitations. None of these theories predicts higher-order mechanisms involving, for example, the interaction of multiple edges (although, if such mechanisms are understood, they can in some cases be treated by the repeated application of the methods described). They all, therefore, fail to predict the scattering at a corner where several edges come together.

They also fail to predict a sometimes important mechanism known as the surface traveling wave. This is an electromagnetic wave initiated when the incident radiation strikes a surface very close to the "grazing angle." The primary surface wave does not directly affect the monostatic RCS because it is traveling away from the source. However, if there is a sharp discontinuity at the far end of a long, thin object, then a reflected surface wave will be generated, radiating a significant amount of energy back toward the source. This occurs in many practical situations, for example in the nose-on RCS of a missile or a long slender aircraft, or of a thin airfoil with a sharp trailing edge. This is one of the reasons why trailing edges, as well as leading edges, must be swept to minimize nose-on signature. A creeping wave is similar to the traveling surface wave for objects that are rounded at the far extremity. In this case, instead of reflecting back from a sharp discontinuity, the creeping wave propagates around the far end of the object, emerges on the other side, and causes a signal to be returned in the direction of the radar receiver.[123]

Effects of surface waves and similar phenomena can be predicted with reasonable accuracy, but these usually require separate calculations in addition to the predictions of scattered radiation by one or more of the methods just described. Calculating the total backscatter from an aircraft usually necessitates applying a composite of the various theories, using the most appropriate method for each major source of backscatter. Furthermore, testing of individual components or features, and eventually of models of an entire configuration, is a vitally necessary complement to analytical prediction techniques in designing low RCS aircraft.

Finally, it must be emphasized that the methods described do not design an aircraft. They only give designers a way to calculate the RCS of some given aircraft configuration. Generating the design to begin with requires the efforts of human engineers and designers. Recently, some computer programs have been developed that are capable of "designing." However, even these programs operate within constraints that are set by the human designers. Extreme care must be taken to ensure that the constraints are appropriate and that the computer is not "suboptimizing," that is, finding a design that appears optimal in view of some specified figures of merit but is not really the best overall solution. Progress must be monitored to ensure that the computer does not generate a design that is unworkable for reasons that may not have been foreseen at the start of the optimization process. In short, the design process will always have to be supervised by humans who are skilled and knowledgable in the entire spectrum of disciplines governing the design and performance of aircraft.

Notation

$\boldsymbol{B}$	=	magnetic induction
c	=	speed of light
$\boldsymbol{E}$	=	electric field intensity
G	=	antenna gain
$\boldsymbol{H}$	=	magnetic field intensity
i	=	pertaining to the incident field (subscript)
J	=	electrical current density
j	=	surface electrical current density
P_{min}	=	minimum detectable received signal power
P_t	=	transmitted power
S	=	energy flux of an electromagnetic wave
s	=	pertaining to the scattered field (subscript)
α	=	angle of incidence relative to a surface normal; or wedge half-angle, depending on context
β	=	angle of incidence relative to an edge
ε_0	=	permittivity of vacuum; $\varepsilon_0 = 8.854 \times 10^{-12}$ F/m
λ	=	wavelength
μ_0	=	permeability of vacuum; $\mu_0 = 4\pi \times 10^{-7}$ H/m
ρ	=	Charge density
σ	=	radar cross section in square meters
σ_{dBsm}	=	radar cross section in decibels relative to a 1 m^2 reference value

Appendix C

The "Skunk Works" Approach

Introduction

Most of the previous discussion has concentrated on management and acquisition strategies employed by the U.S. Air Force and DoD. The unique and not-so-unique features of the prime contractor, Lockheed Advanced Development Projects (LADP), also offer many valuable insights. This chapter therefore describes the approach used by LADP, or the Skunk Works, in the development of the F-117A and other aircraft.

The term Skunk Works, while trademarked by Lockheed, has also come to be commonly used to describe any relatively autonomous special projects, prototyping, and limited production organization within a company, designed to accomplish two key goals:

1) To foster innovations and technological advances that would not be achieved in a more conventional environment.

2) To achieve results more quickly and at lower cost than would occur in a normal environment.

These goals are accomplished by building a small, highly qualified, close-knit team and encouraging creativity and quick, effective problem solving. Designers are encouraged to innovate where necessary and to seek simple, direct solutions where possible. Open and timely communication is substituted for close management oversight. Ben Rich concisely summarized the Skunk Works management philosophy when he became the head of the organization in 1975: "I'll be decisive in telling you what I want, then I'll step out of your way and let you do it. I'll take the crap from the big wheels, but if you screw up I want to hear it first." [197]

Brief History of Skunk Works Projects

The Skunk Works was established in 1943 in response to U.S. Army Air Forces (USAAF) interest in quickly obtaining a jet fighter. Kelly Johnson, who had been at Lockheed for about 10 years and had worked on the Hudson and Constellation transports as well as the P-38 Lightning, was put in charge of the project and given broad authority to accomplish the job the way he wanted. He hand-picked approximately 120 of Lockheed's best engineers and shop people, and disappeared into a screened-off part of a large production building. Johnson reported daily to Lockheed Chief Engineer Hal Hibbard on his progress and needs, as well as to a six-man USAAF project office at Wright Field. The XP-80 prototype (illustrated in Fig. C1) was designed and built in 143 days. Jet engine technology was advancing rapidly, and the XP-80 was almost immediately followed by the larger and more powerful XP-80A, which eventually led to the operational F-80A Shooting Star. With these early projects, Johnson's prototyping organization established not only its reputation but also most of the principles on which it would operate for decades to come.

In 1949 the Skunk Works built the XF-90 (Fig. C2), a "penetration fighter," which lost a competition against the McDonnell XF-88. Lockheed blames the failure of this project on poorly formulated and changing requirements and particularly on designing to a 13.6 *g* ultimate load factor (versus 11 *g*, which was normal at that time), which caused the aircraft to be prohibitively overweight.

Another Skunk Works failure was the XFV-1, a turboprop V/STOL "tail-sitter," which flew in 1954 (Fig. C3). This is claimed to be the only aircraft Lockheed ever built that they were afraid to fly themselves.[106] Engine power surges aggravated the already difficult task of altitude control in hover.

The XF-104 (Fig. C4), which also flew in 1954, was a more successful Skunk Works project. It was a relatively simple aircraft built to a simple requirement. In fact, the requirement was practically dictated to the Air Force by Kelly Johnson himself, with the basic idea that U.S. fighters should be able to fly higher and faster than their adversaries. The aircraft was very fast but limited in other areas such as range, payload, and maneuverability.

Fig. C1 Lockheed XP-80.

Fig. C2 Lockheed XF-90.

Fig. C3 Lockheed XFV-1.

Fig. C4 Lockheed XF-104.

Nevertheless, it became the basis for a series of production F-104 models, some of which were in service with major air forces until very recently.

The next project, the U-2, was designed rapidly during 1954 and made its first flight in 1955. It was built for the CIA and was the first program in which the Skunk Works built and, at least initially, maintained the production aircraft as well as the prototypes. The U-2 (Fig. C5) resembled a sailplane more than a jet in its overall layout, but was otherwise fairly conventional in its design and construction. The only new technology involved was the life support system, needed because of the very high altitudes at which it would be flying.

Lockheed also built the highly successful Agena spacecraft, illustrated in Fig. C6. Although not developed by the Skunk Works, Kelly Johnson was brought in to help structure the design team and the management approach after the pattern of the Skunk Works. Its success has been attributed to a sound design with sufficient safety and performance margins—utilizing proven manufacturing methods and materials—and an innovative but simple passive thermal management system.[76] The XF-104, U-2, and Agena are all examples

Fig. C5 Lockheed U-2.

Fig. C6 Agena in orbit.

of Skunk Works projects that were successful because they were simple, elegant designs that deliberately did not push every aspect of technology.

The Skunk Works also participated in commercial aviation projects. The company designed and built the prototype of the Lockheed JetStar (Fig. C7), which eventually became a fairly successful business jet. However, the classic Skunk Works benefits of team continuity and rapid development were lost as a result of a long gap between prototyping and production.[106]

While the U-2 was making overflights of the Soviet Union, Lockheed and the CIA began planning a high-speed replacement. This first took the form of the liquid-hydrogen–powered CL-400, which the Skunk Works designed but

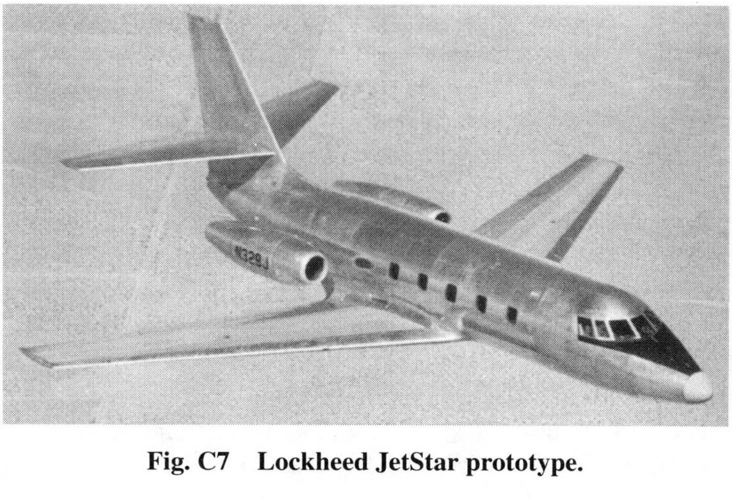

Fig. C7 Lockheed JetStar prototype.

Fig. C8 Lockheed A-12.

quickly recognized as impractical. It was never built. A subsonic, low-signature replacement was also considered; this was referred to as Gusto, but it was not felt that sufficiently low signatures could be achieved to offer lasting protection from the Soviet SAM threat without also utilizing high speed.[175] The real U-2 replacement, the single-seat A-12 (shown in Fig. C8), first flew in 1962. This Mach 3 aircraft "could not have been developed without the use of a prototype."[106] It was an aggressive design, incorporating advances in structures and materials, aerodynamics, flight control, propulsion, and subsystems. An armed interceptor version, the YF-12, was flown and tested but never went into production. The final version, slightly larger and heavier, was the SR-71. The Skunk Works produced the whole fleet and also provided extensive support to the Air Force maintenance depot throughout the SR-71's operational career. There were other Skunk Works projects going on as well, and by 1967 the Skunk Works had grown to 8000 people.[36] However, the traditional Skunk Works management style was reportedly maintained during this period.[197]

One of the other projects going on at this time was the U-2R, an enlarged version of the U-2. This first flew in 1967 and enjoyed considerable success. Another was the D-21, a high-speed (Mach 3.8) reconnaissance drone powered by a Marquardt ramjet similar to the one used on the Boeing Bomarc missile (Fig. C9). The D-21 was designed to be launched from the back of the A-12, but A-12 launches were stopped after a D-21 collided with the launch vehicle on the seventh test flight. Subsequent D-21 flights were launched from a B-52H, using a solid rocket booster to speed up the drone sufficiently for its ramjet to start producing thrust. The program was plagued by problems, and only five operational missions were flown between 1969 and 1973 when the project was canceled.[175]

In the late 1960s and early 1970s, the Skunk Works marketed the Lancer, which was based on the F-104 design with a new engine, wing and tail (Fig. C10). The Lancer lost the Lightweight Fighter design competition in 1972,

Fig. C9 Lockheed D-21 drone carried on modified A-12.

largely because it was not responsive to the goals of the Lightweight Fighter program. The Air Force wanted a new aircraft, designed from the ground up to take advantage of a set of leading-edge technologies, not an old aircraft with new technology "tacked on." The Skunk Works reportedly lost credibility with the Air Force due to Johnson's annoyingly persistent marketing of the Lancer design. The Lancer was never built, although Lockheed photos of a very realistic mockup, built around an actual F-104 forebody, apparently gave some DoD officials the impression that it had been.[13]

Ben Rich became the head of the Skunk Works in 1975, following Johnson's retirement. At this time the Skunk Works' only major aircraft program was the TR-1, another U-2 derivative. Have Blue was the Skunk Works' first new airplane project under Rich and was a great success. It led to the F-117, which followed the pattern of the U-2 and the SR-71 in that the entire fleet was built by the Skunk Works, not just the prototypes.

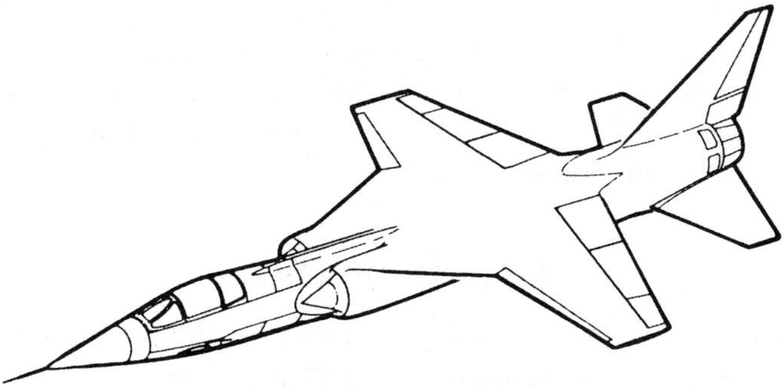

Fig. C10 Lockheed Lancer design.

During the F-117 program, the Skunk Works came under increasing government scrutiny in the areas of technical and financial oversight, security, environmental practices, and safety. Specific requirements imposed on the Skunk Works, broad changes in the acquisition environment, a general consolidation of the aerospace industry and a widespread trend toward teaming on defense projects have fundamentally altered the way the Skunk Works operates. As a result, the F-117 may be the last "classic" Lockheed Skunk Works project.

The Skunk Works became the Lockheed Advanced Development Company (LADC) in 1990. In January 1991, Ben Rich retired and was replaced by Sherman Mullin, who had been Lockheed's F-117A program manager (starting two months before the crash of the first production F-117A). In the early 1990s the Skunk Works moved from Burbank to Palmdale, 35 miles southwest of Edwards Air Force Base. Mullin retired in February 1994 and was replaced by Jack E. Gordon, a veteran of the SR-71, TR-1, and F-117, having joined the Skunk Works in 1964, only about 10 years after Rich. As of 1994, the Skunk Works employed approximately 4500 people.[36]

It is apparent from the preceding discussion that not all Skunk Works projects were successful. This is an inevitable consequence of taking risks, and the whole point of a Skunk Works operation is to encourage innovative and often risky technical solutions that would not be tried in a more conventional environment. "It's taking risks that moves companies and technologies forward," according to Ben Rich.[197] Occasional failures must therefore be considered acceptable. Costs are minimized by *recognizing* failures and dead ends as soon as possible to avoid spending more money on them. Costs are also minimized by keeping programs small until the concepts and/or technologies are proven; technical risk is accommodated by minimizing the financial risk. "A Skunk Works failure is an acceptable research and development risk to top management."[197]

Kelly Johnson noted three important lessons that came out of the Skunk Works experiences[106]:

1) There is no single best method to develop an aircraft.

2) The most common cause of failure is lack of a suitable engine when needed.

3) The second most common cause is working to the wrong [or constantly changing] set of requirements.

Basic Operating Rules

Fairly early in the history of the organization, Johnson wrote down the following 14 basic operating rules for the Skunk Works[124, 187, 197]:

1) The Skunk Works manager must be delegated practically complete control of his program in all aspects. He should report to a division president or higher.

2) Strong but small project offices must be provided by both the military and industry.

3) The number of people having any connection with the project must be restricted in an almost vicious manner. Use a small number of good people.

4) A very simple drawing and release system with great flexibility but also traceability must be provided. [The Skunk Works used a formalized system of "red-lining" for noting changes; new versions of drawings were only issued at certain intervals.]

5) There must be a minimum of reports required, but important work must be recorded thoroughly.

6) There must be a monthly cost review covering not only current status, but also projected costs through program completion. Don't surprise the customer with sudden cost overruns.

7) The contractor must be delegated and must assume more than normal responsibility to get good vendor bids for subcontract on the project. Commercial bid procedures are often better than military ones.

8) Use a simplified inspection system that is agreed to by the customer. Push more basic inspection responsibility back to subcontractors and vendors.

9) The contractor must be delegated the authority to test his final product in flight in the initial stages [of flight test]. Otherwise he will lose his competency to design other vehicles.

10) The specifications applying to the hardware must be agreed to in advance of contracting. The ADP practice of having a specification section stating clearly which important military specification items will be knowingly not complied with, and why, is highly recommended.

11) Funding for a program must be timely so that the contractor doesn't have to keep running to the bank to support government projects.

12) There must be mutual trust between the military project organization and the contractor, with very close cooperation and liaison on a day-to-day basis.

13) Access by outsiders to the project and its personnel must be strictly controlled by appropriate security measures. This is necessary to maintain productivity even if no security requirements exist.

14) Because only a few people will be used in engineering and most other areas, ways must be provided to reward good performance by pay not based on the number of personnel supervised.

The importance of good recordkeeping, *particularly* when formal documentation requirements are relaxed, is worth emphasizing. The Skunk Works was very thorough in this respect. According to Lt Col Randy Smith, who was recently a member of the 2762nd Logistics Squadron (Special) at Norton Air Force Base, California, the principal Air Force support facility for both the SR-71 and the F-117A,

> All design reports were meticulous—phenomenal, in my opinion! They...did an excellent job of documenting all aspects of design, including spec noncompliance in design reports; i.e. MIL SPEC requirements to design for 2 *g* taxi loads were relaxed to 1.8 *g* since 2 *g* is like 1×10^{-6} probability" per flight, which is extremely unlikely to occur even once over the life of a 30-aircraft fleet. Although documentation requirements were relaxed in these "Special" programs, compliance or noncompliance with relevant specifications was still carefully recorded, typically neatly handwritten by the engineers who had done the work originally.[216]

Other Skunk Works principles include the following:

1) Engineers and designers should be "within a stone's throw" of the shop floor.[258] This allows design and manufacturing problems to be worked promptly and interactively between the designers and the machinists, and in the long run it causes the designers to become more familiar with the manufacturing process. Engineers of different disciplines also need to be located close to each other to facilitate rapid solution of mutual problems. This created a "systems engineering" environment long before the phrase was even coined.

2) Bad news must travel to the top at least as quickly as good news. Individuals are encouraged to pursue innovative approaches, but when there's a problem then the person who first becomes aware of the problem is expected to report it promptly.

3) Keep it simple. Except where necessary to achieve specific capabilities, a system should be as uncomplicated as possible. Examples of Skunk Works projects that were successful in part because of their relative simplicity include the XF-104 fighter prototype, the U-2, and the Agena spacecraft, which, although not developed at the Skunk Works itself, was very deliberately developed using a Skunk Works approach.

COMMENTS AND OBSERVATIONS

Any organization can set these priorities and rules. That will not guarantee success of a particular project or of the organization as a whole. A combination of other factors worked together for the benefit of the Skunk Works. These include personal leadership, strong support from the government, organizational momentum, and a solid reputation based on a long history of accomplishments.

PERSONAL LEADERSHIP

Kelly Johnson was the head of the Skunk Works for more than half of its existence to date. He was a technical leader as well as a manager, having come from an era when the aircraft designer was expected to be a master of nearly every discipline. He had personally trained much of the Skunk Works workforce and frequently made engineering decisions himself when problems were brought to his attention. In fact, the Skunk Works basic groundrules, which he wrote, specifically gave him the ultimate decision authority in all matters. As a manager, his personality motivated people to deliver their best efforts. He was always looking over people's shoulders and asking questions, effectively "red-teaming" everybody who worked for him and making sure they had good reasons for whatever they were doing. If he was satisfied, he trusted and supported his employees, encouraged creativity, and rewarded accomplishments. But he could also be abusive if an employee failed to perform or violated Skunk Works ethics. With his notorious temper, Kelly Johnson might not be considered a good candidate for management in today's workplace. But his leadership was unquestionably effective, and it was a crucial element to the success of the organization that he built.

Changes occurred in 1975 when Ben Rich became head. He had been at the Skunk Works since 1954 and worked on propulsion integration for the U-2, A-12, and SR-71. Although he was a highly capable engineer in his field (thermodynamics/propulsion), Rich did not have Johnson's breadth of technical expertise, and he did not try to fill Johnson's shoes as a technical leader. More responsibility therefore rested upon the employees. Kelly's presence was still felt. He was retained as a consultant, and of course most of the senior Skunk Works people were accustomed to working in the ways that Kelly had established. Nevertheless, Rich was in charge from 1975 on, and he deserves much of the credit for enforcing, in effect, a "low observables culture" at the Skunk Works, in which the new discipline of signature control was made coequal (or even preeminent) to aerodynamics, structures, propulsion, and so on.[197] This cultural change was crucial to the successful development of a very low observable aircraft. The RCS improvements that were sought simply could not have been achieved without full cooperation in every engineering discipline.

GOVERNMENT ASPECTS

Regardless of a company's internal policies, strong support from the government is necessary to successfully implement a streamlined, aggressive approach from the contractor. The government agency responsible for developing and acquiring a new weapon system from a Skunk Works–like organization must respect and observe the principles outlined in Kelly Johnson's 14 basic operating rules.

INSTITUTIONAL MOMENTUM AND REPUTATION

The Lockheed Skunk Works had a special relationship with its customers, largely the U.S. Air Force and the CIA. When needs arose, these customers often bypassed standard competitive procurement methods and went straight to the Skunk Works, justifying this on the basis of the Skunk Works' unique capabilities (this was not done illegally, but rather only with proper justification and approvals for sole source decisions). Once contracts were awarded, they gave the Skunk Works unusual leeway to accomplishing tasks. These special considerations were given because of the Skunk Works' past history of accomplishments, which also helped assure support for the Skunk Works from Lockheed corporate management. Although the Skunk Works had its share of failures, those tended to fade into obscurity, whereas the successes made a lasting impact. The value of the unique national resources generated by the Skunk Works far outweighed the money spent on all projects including the unsuccessful ones.

Other Examples and Common Factors

Although Lockheed was undoubtedly the most successful at applying the Skunk Works approach to many projects over an extended period of time, other companies did at times create the same kind of environment to accomplish rapid prototyping of innovative designs.

One example is the North American P-51 Mustang, illustrated in Fig. C11. In late 1939, the British Purchasing Commission was looking for a manufacturer to build Curtiss P-40s under license, because all of Curtiss' projected output of P-40s was already committed. North American proposed a new fighter, and in a letter contract dated April 11, 1940, the British placed an order for 400 of the new aircraft, specifying that the first delivery should take place in January 1941 and be completed by September 30, 1941. North American Vice President Lee Atwood took personal responsibility for meeting these deadlines. The prototype made its first flight on October 26, 1940, just six months after the letter contract was received from the British.[51] The Mustang incorporated a variety of improvements ranging from aerodynamics to

Fig. C11 North American P-51 Mustang.

producibility. The airframe was manufactured in five sections, each of which could be essentially completed before they were brought together for final assembly. The distinctive squared-off wing and tail planforms were also a concession to producibility. The designers realized that rounded wingtips, popular in the 1930s, offered little or no real aerodynamic benefit. Partly as a result of these features, the P-51 production rate at North American's Inglewood, California, factory was able to reach 22 *per day* in 1944.[11] The airfoil used on the Mustang was a new "laminar flow" section, and although in practice it probably did not achieve significant laminar flow, it nevertheless offered reduced drag relative to conventional airfoils of the time. A tank in the fuselage just aft of the cockpit, in addition to the internal wing tanks, provided a large internal fuel capacity, which could be further augmented by external tanks. With the aft tank full, the P-51 had very poor handling characteristics with little or no longitudinal stability; so the aft fuel had to be burned prior to combat.[101] Nevertheless, the large fuel capacity coupled with the low drag of the P-51 provided exceptional range. The P-51 was the only fighter that could escort Allied bombers all the way to Berlin and this fact alone had a significant impact on the war. Yet, at the time the P-51 was developed, the U.S. Army was working under a cumbersome acquisition process that, combined with a lack of high-level official interest, inhibited the development of innovative, high-performance aircraft.[11] North American took advantage of a set of circumstances, in particular foreign interest and the urgency of impending world war, to step outside the normal acquisition process, accept some risk, and turn loose its designers. The result has gone down in history as one of the finest fighters of World War II.

A contemporary of the Mustang, also developed by a small, innovative team, was the Focke-Wulf 190, shown in Fig. C12. The Fw 190 was designed by 12 engineers led by Kurt Tank, Focke-Wulf technical director. Three of

Fig. C12 Focke-Wulf (Fw) 190.

the designers were also experienced test pilots who participated in the Fw 190's test program. First flight was accomplished within one year of the German Air Ministry's initial request for proposal. The designers' aim was to develop a relatively simple fighter that would be more supportable than other state-of-the-art fighters. They stressed producibility, ruggedness, mission flexibility, simplicity of operation, and carefree flying qualities, rather than just absolute speed, which had been the main emphasis in fighter development up to that time. They achieved their aim by using an air-cooled engine that eliminated the complexity and vulnerability of a radiator system; electrical rather than hydraulic landing gear and flap actuation; and a modular, removable front end that allowed relatively easy replacement of the engine or the oil cooler assembly. For reduced pilot workload and improved situational awareness, the Fw 190 featured superbly harmonized flight controls with light stick forces; a bubble canopy that most other fighters only added as a modification later in the war; a reclined seat for improved *g*-tolerance; and a unique integrated propulsion control system that relieved the pilot from worrying about the proper relationship between altitude, propeller pitch, engine boost, mixture, and RPM, as fighter pilots typically had to do. When introduced into combat in late 1941, the Fw 190 established clear and immediate ascendancy over contemporary versions of the Spitfire. As the war progressed, the flexibility inherent in the 190 allowed it to perform a variety of missions and, with upgrades, to remain equal to the best Allied fighters developed. In its ultimate form as a high-altitude interceptor, with extended wings and a new engine, it was able to operate at altitudes of more than 45,000 ft and at speeds of more than 470 mph, making it the highest-flying and one of the fastest piston-engined fighters of the war.[191]

The P-51 and the Fw 190 are both examples of what can be accomplished when a small team with clear goals is free to innovate and succeed.

Another example is the General Dynamics YF-16 Lightweight Fighter. This development effort is well documented and bears striking similarities to the Skunk Works approach. The Lightweight Fighter program of 1972 was part of a new Air Force streamlined prototyping effort, initiated in response to criticisms of government micromanagement, excessively long development times, and large expenditures on unproven systems during the 1960s. Air Force and DoD leadership were therefore committed to fostering an atmosphere in which the contractors could exercise creativity and innovation with minimal interference. The program was provided with strong support, and traditional government requirements were relaxed. According to Harry Hillaker, who was deputy chief engineer for the YF-16,

> The contractor, General Dynamics, likewise, departed from his normal organizational structure and operating procedures to establish a distinctly separate prototype organization....The team was headed by a vice

> president and program director who reported directly to the president of the Convair Division (consisting of the Fort Worth Operations and San Diego Operations) who, in turn, reported to the chairman and CEO of General Dynamics. His authority (the YF-16 program director) bypassed the general manager of the Fort Worth Operations and the corporate chief operating officer. He had the authority to make any corporate commitments to the Air Force on the YF-16 program (except the acquisition of new plant equipment or facilities) if the situation demanded such....The total YF-16 team reported directly to the program director. That team, with all its functions was collocated adjacent to the factory area where the airplanes were fabricated, assembled, and tested—communications were direct and paper work was reduced to a minimum. All discrepancies, interpretations, and the like, were addressed immediately and face-to-face. In most instances problems were resolved before the beginning of the next working day.[102]

Hillaker also emphasized the importance of strong but flexible leadership:

> The success of a direct, flexible management style is quite dependent on the type of person selected to manage the various elements of the program. It must be someone who is *result oriented* and *not process or procedure oriented*....The importance of a process is not that it be followed blindly, but that it shapes rational thinking or analysis.[102]

Whatever procedures an organization adopts, it is the results that matter. Rigid adherence to any set of policies or procedures can stifle innovation, add "busy work," and generally block the very same progress that the procedures were intended to encourage. "Process is what you do to vegetables, not how you run a program." [103] The YF-16 team and the Skunk Works share many common features:

1) Separate entity from the rest of the company. This helps focus effort and minimizes distractions, which is an important benefit of a Skunk Works type of organization. Day-to-day distractions and interruptions can have a crippling effect on the productivity of even the most capable workers.

2) Design team collocated with the factory area. Engineers communicate directly with shop workers, take "cradle-to-grave" responsibility for what they design. Over time, this gives designers a better understanding of manufacturing techniques and what the cost drivers are.

3) Strong, capable leadership with broad authority and relatively direct access to top corporate decision makers, and frequent and direct contact with working level people within the organization ["management by walking around" —Jack Twigg, Col U.S. Air Force (Ret)],[234] This gives the director a good understanding of the problems his or her people are dealing with, and the ability to rapidly bring corporate resources to bear to solve those problems.

4) Internal workings not overly rigid. Face-to-face, near-real-time decision making. Minimal formal documentation, but each individual was expected to keep records of his or her work.

5) Frequent, open communication with customer to avoid surprises. Frequent but informal design reviews, with few participants—only those directly involved in a particular aspect of the project.

CONCLUSIONS

Whether at Lockheed or at any other company, the objectives and many of the methods are the same. The idea is to establish a distraction-free environment of teamwork, creativity, and appropriate risk-taking, not bound by conventional practices, to deliver direct, sensible technical solutions quickly and at moderate cost. These goals are best accomplished when both the government and the contractor maintain small organizations: "Workers and managers alike can deal with individuals, not departments, and problems can be resolved in a matter of days rather than weeks or months." [236] According to Ben Rich,

> The successful program will involve a small team working together rather than the large bureaucratic hierarchy. This is in part a function of the fewer relationships and the opportunity to build trust, as opposed to the adversarial and more distant arms length relationships involved in the large hierarchical program.... Decision making is also quicker in a small organization than a large one. The bureaucracy favors avoiding decisions and taking absolutely no risk.[196]

These ideas are well known. However, serious disconnects between acquisition policies and visions, and the procedures actually implemented, have created an environment that is not conducive except in special cases to true Skunk Works operations. The late Ben Rich noted, "As time goes on it becomes increasingly more difficult to maintain the Skunk Works attitude.... New legislative initiatives become more difficult to resist. As the program becomes older, security requirements are relaxed and outside 'help' becomes harder to resist. Those who are not permitted in become convinced that something wrong must be going on...." [196]

Programs tend to grow with time, and so the ability of a program to stay small is tied to its ability to reach completion as rapidly as possible. The tendency to trim near-term budgets by stretching program schedules works against this. Teaming and subcontracting also make it more challenging to achieve the tight-knit organization and face-to-face real-time decision making that is an important element of Skunk Works problem solving. As Ben Rich aptly stated,

> I don't honestly believe that you can take the Skunk Works approach and bottle it up and sell it.... I do believe that you can apply these basic principles in select acquisitions.... But it does require a willingness to create a small team and keep it small. To choose the right people, to give them clear stable requirements, and a job to do. To be willing to let them alone to do that job and to provide them with stable funding.[196]

Factors that are conducive to Skunk Works methods include "applied early in the acquisition, all parties agree that time is of the essence, to prove or apply a technical breakthrough."[196] The Skunk Works approach is not right for every program, but in certain situations it would be very beneficial for the government to give a company the freedom to operate in this manner. Not only the agency directly involved with the acquisition but also the top leadership of the military service that is involved, together with DoD and even Congress, must cooperate and be willing to relinquish some of the usual control and oversight in exchange for the opportunity to achieve a technological breakthrough or field a revolutionary new weapon system in a rapid, cost-effective manner.

Appendix D

Contemporary Reports on Early Low Observables Development

A brief reference to the DARPA-sponsored "stealth fighter" program was made in an article on advanced tactical fighter studies in *Aviation Week and Space Technology (AW&ST),* Jan. 19, 1976, just two months before the Lockheed/Northrop "pole-off":

> Air Force will initiate a series of conceptual design studies of an advanced tactical fighter (ATF) this spring....
>
> Besides the overall design studies, the Flight Dynamics Laboratory intends to sponsor two other ATF studies in the area now generally called Stealth—ways and means of reducing an aircraft's radar, acoustic, and optical signatures.... Flight Dynamics Laboratory had been asked to use Northrop and Lockheed as its Stealth contractors because these two companies have been working on a Stealth fighter aircraft program sponsored by Defense Advanced Research Projects Agency. But the laboratory is planning to compete the two ATF Stealth awards.
>
> The intent of the Stealth studies will be to explore how far it is practical to progress in the application of signature reduction techniques such as aerodynamic body shaping, absorptive materials, and paint coverings without compromising the system's operational capability. They also will seek to identify how much of the remaining signature could be compensated by electronic countermeasures.[174] *[Note that this article did not mention infrared signature reduction and avionics emissions reduction.]*

Aerospace Daily carried a one-and-a-half page article in July 1976 of which excerpts follow:

> ***Lockheed's Kelly Johnson Building "Stealth" Aircraft***—The nation's premier aircraft designer has returned to the famous "Skunk Works" at Lockheed, where he is building a new 12,000-lb, one-man "stealth" aircraft. It is due to fly in about two years.
>
> The program is sponsored by the Defense Advanced Research Projects Agency (DARPA), and sources tell The DAILY it will cost about $90 million over the three-year development period.

> The goal is to reduce aircraft visibility—optical, acoustic, and radar signatures—through new technology *[as in the previous article, infrared signature and active avionics emissions reductions are not mentioned].*
>
> Driving force behind the stealth aircraft project is Ken Perko of DARPA, who was formerly with the drone system program office at the Air Force's Aeronautical Systems Div.
>
> After original studies were completed about a year and a half ago [late 1974 or early 1975] by McDonnell Douglas, Lockheed, and Fairchild Industries, Perko sent up the line a proposal to build one or two demonstration aircraft to be funded jointly by the Air Force, Navy and DARPA. Each would have put in about $10 million a year for three years.
>
> The idea reached Air Force Chief of Staff Gen David Jones and Chief of Naval Operations Adm James Holloway, but was not accepted. DARPA then decided to proceed on its own about a year ago [July or August 1975]. The Air Force Flight Dynamics Laboratory, meanwhile, is continuing its own Air-to-Surface Technology Evaluation Integration (ATS) program, in which stealth is a major concern (DAILY, July 20).
>
> The Air Force has been interested in characteristics applicable to a stealth aircraft for some time. In the fiscal year 1972 budget, for example, it proposed prototyping of an unmanned very low radar cross section test vehicle, and a quiet aircraft. The projects, turned down by Congress, were part of a prototyping package that included the Lightweight Fighter [LWF] and the Advanced Medium STOL Transport [AMST], both of which were approved. *[The prototyping package referred to was proposed by an Air Force Prototype Study Group, which was formed in 1972 under a DoD-wide Advanced Prototype Development Program. The LWF Program led to the YF-16 and YF-17, while the AMST Program produced the YC-14 and YC-15. Source: Aronstein and Piccirillo.]*
>
> Neither of the two projects was aimed at specific applications, but they were chosen [by the Air Force Prototype Study Group] over many others as well worth demonstrating.
>
> The low radar cross section vehicle would have been "below the detection range of most radars," one official told The DAILY at the time. Speed was not seen as a big factor, and stress was to have been on cutting observables, with planform design and materials exploitation seen as best routes to the objective....[131]

There were other references in *AW&ST* and Jane's *All the World's Aircraft* to the Have Blue program (although not by name) through mid-1977, and these were short but mostly accurate:

> Lockheed California Co. is developing a small fighter intended to demonstrate stealth, or low signature, technologies under contract from

> Air Force Flight Dynamics Laboratory, funded by Defense Advanced Research Projects Agency. The company won the contract in competition with Northrop Corporation, after both had earlier DARPA studies.[15]

> Lear Siegler is readying an advanced digital flight control system for the small stealth fighter being developed by Lockheed-California under contract from Air Force Flight Dynamics Laboratory with Defense Advanced Research Projects Agency funding (*AW&ST,* Aug. 2, 1976, p. 11). The stealth fighter is intended to demonstrate the low signatures required by advanced fighters in an environment of hostile radar, infrared and other sensors.[16]

> First flight of Lockheed's new Stealth Fighter demonstrator being built by the company's "Skunk Works," where retired director C. L. "Kelly" Johnson is now a consultant, will be conducted late this year. The aircraft, powered by a pair of General Electric J85 turbojet engines, is intended to demonstrate low radar, infrared and optical signatures to enhance survivability in hostile environments (*AW&ST*, Feb. 14, p. 9). The project is funded by the Defense Advanced Research Projects Agency and administered by the Air Force.[17]

> ***Lockheed "Stealth Fighter"***—In the so-called "Skunk Works" at Burbank, Lockheed-California is reported to be building, under an ARPA-funded contract from the USAF's Flight Dynamics Laboratory, a small "stealth fighter" of which a primary feature will be low radar, infra-red and optical signatures. The fighter is said to be powered by two General Electric J85 turbojet engines, and was expected to fly before the end of 1977.[104]

The two editions of Jane's (1978–1979 and 1979–1980) that followed carried a slightly "updated" version of the same entry, with the reported engine size increased to approximately that of the F117-A rather than Have Blue:

> ***Lockheed "Project Stealth"***—In the so-called "Skunk Works" at Burbank, Lockheed-California is reported to be building, under an ARPA-funded contract from the USAF's Flight Dynamics Laboratory, a single-seat reconnaissance/strike aircraft, of which a primary feature will be low radar, infrared, and optical signatures. The aircraft is said to be powered by two 53.4 kN (12,000 lb st) turbojet engines and is believed to have flown for the first time in 1977. [104]

The next new references to stealth technology appear to have been in the 1979–1980 timeframe, when several *AW&ST* reports focused on the advanced technology bomber and the political maneuverings surrounding the 1980 election and the possible resurrection of the B-1:

> ***Advanced Bomber, Missile in Definition***—U.S. Air Force is in the earliest stages of defining a new strategic manned bomber.... The Rockwell International B-1 earlier terminated by President Jimmy Carter and the General Dynamics FB-111H re-engined version of the operational FB-111A, both terrain-following bombers, will be considered along with more-advanced designs in the new manned bomber program.... Studies show that supersonic speed, at least in a low-level environment, may not add comensurately more survivability when compared to bomber defense systems and low radar cross-section techniques....This is in advance of the formal U.S. Air Force new manned bomber mission element needs statement planned soon and a U.S. Air Force program management directive which will give formal approval for the development effort.[41]
>
> ***Multipurpose Bomber Advances***—Defense Dept. and the House-Senate Authorization Conference Committee walked a tightrope in seeking a compromise solution that the White House will accept in determining the Fiscal 1981 funding for a new multipurpose bomber.... "There was strong sentiment within the conference committee that perhaps what we ought to have is a mixed force with 100 of the more conventional bombers along the lines of a modified B-1, and 100 bombers coming along later using all new technology, particularly stealth qualities to avoid radar detection," one House staff member said.[199]
>
> ***Bomber Biases***—As long as the options remain open for the stretched General Dynamics FB-111, the B-1 and the advanced technology "stealth" bomber, the White House intends to continue studies. Some Administration officials believe a delay in the studies would allow more time to perfect stealth technology....[18]
>
> ***New Bomber***—Stealth low-radar-observability technology could be incorporated easily in the new manned bomber Congress has ordered to be built by 1987.... [Senator Alan] Cranston, co-author of the Glenn–Cranston amendment to the Fiscal 1981 Defense Authorization Bill, said he knew of stealth technology when the amendment was written, calling for selection by Mar. 15, 1981...of a stretched General Dynamics FB-111, Rockwell B-1 version or an "advanced technology" bomber. The 1987 date [for initial deliveries] was chosen..."precisely to allow the Secretary of Defense [Harold Brown] to consider the stealth technology program," Cranston said.[19]
>
> ***Stealth Aircraft Leaks Deplored***—Stealth low-observability technology (*AW&ST* Sept. 1, p. 41) should have been discredited by the Defense

Dept. when the first news leaks about the program started to appear, Gen Richard H. Ellis, commander of the Strategic Air Command, said...

Several [House Armed Services investigations subcommittee] members criticized Defense Secretary Harold Brown and his subordinates for overreacting to the leaks by releasing too much information on the program.

Brown said public announcements were made along with the establishment of a new security system.... The Defense secretary added that he chose to acknowledge what the Soviet Union had learned through the leaks and to lay down strict new security guidelines to prevent further leaks. He said that option was chosen over two other possibilities—a series of "no comments," which would have led to new leaks and further competitive reporting, or discrediting the story, a public deception that Brown said would undermine "the basic link of trust between the government and the people...."

William J. Perry, under secretary of defense for research and engineering... said that in 1977 Stealth had reached the technology demonstration phase, at which point the decision was made to increase technology development, accelerate applications of the technology and place the whole effort in a separate security compartment. Since then, Perry said, he has had responsibility for the program.

"In 1977," he said, "I told the secretary that with good luck we could conceal program existence for two years. In fact, we have kept its very existence secret for more than three years." [20]

OSD was able to acknowledge the existence of a stealth technology program, and honestly report that it had reached the demonstration phase in 1977, without actually saying anything about the F-117A, which was already nearly two years into its full-scale development and less than one year from first flight. The bomber story, of course, culminated in the B-1B and B-2 programs. On the same day as the last article quoted, there was a reference in *AW&ST* to "Project Harvey," a name that has commonly (but mistakenly) been associated with the DARPA studies that led up to Have Blue:

Rabbit Ears—A Pentagon leak monger who wears three-piece suits that perfectly match the GI paint decor of the fabled E-ring faded into view last week with this: 'You didn't hear it from me, but we don't use Stealth anymore. It now is Project Harvey, and we are thinking of recalling Brig Gen Jimmy Stewart to head it, if he can get away from Hollywood for a while. Keep me out of it.' And he hopped off down the E-ring until he merged back into the paint again.[20]

Nothing related to Have Blue or the F-117A was ever called "Project Harvey," according to Jack Twigg, who as a major in the Air Force was the Have Blue program manager. "All activities that led directly to Have Blue were *never* referred to as Project Harvey. There was a gentleman in the Pentagon...who had some concepts that he entitled 'Harvey.' His focus was tactics and other means, *not* 'technical' low observables.... The occasional mention of 'Harvey' did provide confusion to some, and *may* have served as an inadvertent cover story."[234] Adding to the confusion is the fact that both "Harvey" and Have Blue drew support from within the Office of the Director of Defense Research and Engineering (ODDR&E) at the Pentagon, making the distinction very unclear even to many program participants. Note that this article appeared in late 1980, well *after* the completion of Have Blue.

Just over a year later, some information on the "stealth fighter" did come out in *AW&ST:*

> ***Stealth Projects***—Defense Dept.'s $21.9-billion advanced technology bomber—the Stealth aircraft—is one of three low radar profile technology programs....The Lockheed Stealth fighter, approximately the size of the Navy/Northrop/McDonnell Douglas F-18, will fly this year.... [the F-117A had actually flown in June]. The initial lot is for 20 of the special mission aircraft, which will be based in the southwest desert and flown by Lockheed C-5 transports to areas of the world where needed. The Stealth fighter resembles in design the NASA space shuttle's planform....
>
> In the Defense Advanced Research Projects Agency (DARPA) Have Blue development program for a subscale fighter demonstrator aircraft, which led to the Lockheed fighter program, two of the Stealth aircraft crashed. Both mishaps were unrelated to Stealth technology. All 20 Stealth fighters will be in the inventory within two years.[21]

This report was fairly accurate in most matters of history, dates, and program plans. Limited initial operational capability was, in fact, achieved almost exactly two years after this article, although only 14 of the aircraft were in the inventory by that time. The statement about the aircraft's shape, on the other hand, was completely inaccurate.

Little, or no additional, correct information appeared during the early- to mid-1980s. Lacking facts, the popular aviation press continued to speculate as to the design of the "stealth fighter," also referred to as the "F-19." It was pictured as an elegantly curved, highly supersonic machine resembling the SR-71.

The extreme security measures at the crash site following the loss of an F-117A in July 1986, however, left the public in little doubt as to the *existence* of the "stealth fighter." The *Associated Press* reported:

> Air Force guards with automatic rifles sealed off a wide area of a national forest Saturday around the burned crash site of a mysterious aircraft widely speculated to have been an experimental Stealth fighter.

> The Air Force imposed strict secrecy and said little Saturday other than to identify the pilot and say he was stationed at Nellis Air Force Base in Nevada.... Lt Col John Kuminecz at Nellis said the pilot was attached to the 4450 Tactical Group at Nellis, and added: "I can't comment on the mission of that group."
>
> Col Albert C. Piccirillo, head of the Advanced Tactical Fighter Program office at Wright–Patterson Air Force Base in Dayton, said all he knew about Friday's crash...was what was in the local newspaper.... Military officials refused to divulge what kind of plane crashed...but numerous published reports said the plane apparently was an F-19 Stealth fighter or a prototype for it....
>
> Though the Pentagon has acknowledged that Stealth technology is being used by Northrop Corporation to develop a bomber, both the Air Force and Lockheed have consistently refused to respond to queries about a fighter with such characteristics....[178]

Another article about six weeks later in *The Washington Post* contained further information about the program, as well as the first reasonable information about the airplane itself:

> The United States has about 50 radar-eluding "Stealth" jet fighters fully operational under a top secret multibillion-dollar program, informed sources said yesterday.
>
> Several squadrons of Stealth fighters are hidden in hangars in the Nevada desert near Tonopah, southeast of Reno, to avoid detection. The planes fly at night there under the control of the Air Force's Tactical Air Command, sources said. The mysterious plane that crashed near Bakersfield, California, on the night of July 11 was a Stealth fighter on a training mission, sources said.
>
> The Air Force, which originally planned to buy 100 Stealth fighters, is expected to settle for about half that number because of high costs and other budget demands, according to one knowledgeable source.
>
> The cost of the Stealth fighter program is in dispute. One source estimated that it will total $7 billion *[fairly accurate]*, and that the cost per plane already exceeds $100 million, including development costs. But other Defense Department officials countered that such an estimate is too high. The F-15, the most sophisticated fighter now in public view, costs about $40 million each *[the dispute results from the fact that a comparison is being made between the total program cost per aircraft of the "stealth fighter" and the unit flyaway cost (UFC) of the F-15. UFC does not include research and development or certain nonrecurring procurement costs. In fact the total program cost of the F-117A does come out to just over $100 million per aircraft, but its UFC is $42.6 million].*
>
> The Air Force still refuses to acknowledge the existence of the Stealth fighter....

> The Stealth fighter is a derivative of the SR-71 Blackbird, also developed in Lockheed's "Skunk Works" under famed aircraft designer Clarence L. (Kelly) Johnson, sources said. But compared to the sleek Blackbird the Stealth fighter looks "ugly" because of its bulging, nontraditional shape, according to several sources who have seen it.
>
> Nevertheless, Air Force officers are excited about the Stealth fighter's demonstrated ability to elude detection by radar and infrared systems....
>
> The Stealth fighter has one basic purpose, sources said: to sneak up on a target at relatively slow speed, launch a missile or "smart" bomb, and run home before the enemy realizes what happened. The Stealth fighter is not as fast or agile as the supersonic F-15 fighter....
>
> Successors to this first generation Stealth fighter will have much better performance in terms of speed, range, altitude and maneuverability while retaining [stealth], the sources added.
>
> The recent public disclosures about the Lockheed Stealth fighter have prompted Defense Department officials to discuss privately how the existence of the aircraft might be officially revealed....[254]

This revelation was not made until November 1988, at which time the program was officially acknowledged and a few deliberately poor-quality photographs were released. The public did not get a *good* look at the F-117A until its first public showing in April 1990. Almost immediately thereafter, a spate of books and articles described the F-117A in detail, complete with close-up photographs of many of its unique features.

Appendix E

Senior Trend Acquisition Plan, July 1980[241]

Acquisition Plan No. RWSX-80-3

Program: Advanced Technology Aircraft (Senior Trend)

Description of System: The Advanced Technology Aircraft is a low observable fighter to be used as a penetrator in tactical situations. This production program includes 20 aircraft, 60 General Electric F404/F1D2 nonafterburning turbofan jet engines, aeronautical support equipment, training, contractor support, initial spares and repair parts, engine containers, and associated data.

Approved:

[signed] JERRY W. BABER, LT COL, U.S. Air Force — July 31, 1980
Contracts Representative

[signed] ELDRED D. MERKL, COL, U.S. Air Force — July 31, 1980
Program Manager

[signed] DAVID B. ENGLUND, COL, U.S. Air Force — July 31, 1980
System Program Director

DESCRIPTION OF SYSTEM

The Advanced Technology Aircraft (ATA) exploits recent breakthroughs in low observables technology to negate anticipated (1990) Soviet defenses. The ATA will be employed in offensive operations and in a manner that contributes to increased effectiveness of other friendly weapon systems. The aircraft will be operated to capitalize on the aircraft's low observable signatures to confuse, disrupt, and destroy the enemy's war making capability. The primary mission will be to attack enemy surface targets with conventional weapons. The principal targets will be command and control communications centers, air defense facilities, airfields, logistics choke points, and other targets of high military value. The ATA will be powered by the General Electric F404/F1D2 engine, a nonafterburning derivative of the U.S. Navy F404-GE-400 afterburning turbofan engine. The F404/400 is currently the power plant for the F-18A/A-18A aircraft. The F404/F1D2 retains maximum commonality with the F404/400, deviating only where necessary to provide the nonafterburning configuration and the specific systems requirements of the ATA.

The Air Force has established the program priorities to be security, observables, schedule, other system performance, and cost. While security remains the absolute highest program priority, the ATA project is structured to meet observable requirements without degradation of aircraft performance, and at a reasonable cost to the government. Special management and reporting arrangements have been established involving face-to-face communications on a regular basis among various Government agencies and between the program office and the contractor. Formal documentation will be the minimum essential for contracting, program and financial planning and reporting, design, and interface definition, testing, maintenance, and support.

PROGRAM FUNDING (PRODUCTION)

Program funding is provided from more than one budget line item and program element to enhance the overall security of the program. The following total program funding plan is based on contractor estimates and program office analysis.

20 A/C + 60 Engines—TY $ in Millions

Fiscal year	*Prior*	*80*	*81*	*82*	*Total*
Required program funding	XX	XX	XX	XX	1025.1

The planned production buy supported by this funding profile is as follows:

Fiscal year	*80*	*81*	*82*
Aircraft	7	8	5
Engine	14	22	24

Long-lead funding is contemplated for the FY 82 effort by this acquisition plan. This will allow the acquisition of those long-lead time ATA components, which are necessary to protect the out year delivery schedule.

Delivery Requirements

Schedule is the third highest priority in the program with planning designed to achieve the earliest initial operational capability (IOC) within known fiscal and technological constraints. The risk involved in the concurrency of development and production is recognized and accepted as commensurate with overall program objectives. The following is a breakout of aircraft and engine deliveries that will meet current program requirements.

Background and Procurement History

The Senior Trend airframe full-scale development effort was initiated by CPFF Contract F33657-79-C-0016. The General Electric F404/F1D2 engine full-scale development effort was initiated by CPFF Contract F33657-79-C-0017. The flight test portion of the full-scale development is scheduled to be completed in the second quarter of FY 83.

Program Risk—Technical and Schedule

Schedule concurrency is necessary to achieve a July 1982 initial operational capability. Therefore, the Senior Trend program is structured with a degree of technical and schedule risk. The program office has evaluated and will continue to evaluate each risk area in terms of necessity, acceptability, and parallel efforts to reduce risk.

Integrated Logistics Support Planning Concept

Two levels of maintenance, field/organizational (F&O) and depot, will be used to support the ATA. F&O level maintenance will be limited to the removal/replacement of line replaceable units and shop replaceable units. F&O maintenance will be performed at the home station, main operating base, and forward operating locations by the using command. depot level maintenance will be performed by the contractors of the ATA, engines, and associated subsystems.

Approval for Operational Use

In addition to satisfactory completion of full-scale development and initial operational testing of the ATA, approval must be obtained from DoD prior to operational use.

Identification of Participants in the Acquisition Preparation

Col E. D. Merkl	Program Manager
Lt Col Jerry W. Baber	Business Manager
Capt Dennis Ferry	Budget Analyst
Capt Clinton J. Asbury	Contracting Officer

Source

The contractors for this acquisition are:

General Electric Company	Lockheed Corporation
Aircraft Engine Group	(Lockheed California Company)
1000 Western Avenue	P.O. Box 551
West Lynn, MA 01910	Burbank, CA 91520

The Lockheed Corporation was originally selected for the development of the ATA because of its demonstrated capability in low observables. The General Electric F404/F1D2 engine (a derivative of the F-18 F404-GE-400 engine) was selected as the ATA propulsion system based on compatibility, performance, schedule, and cost.

Determination and Findings

The enclosed Determination and Findings covers the production effort for the ATA Airframe and the F404-GE-F1D2 engine to be acquired as government furnished equipment.

Contract Type

It is contemplated that fixed price incentive firm (FPIF) contracts will be negotiated for the airframe and engine production efforts.

Negotiation Authority Recommended

Negotiations pursuant to the authority of 10 U.S.C. 2304(a)(14) will be used because the manufacturers have been substantially funded by the government for the aircraft and engine full-scale development efforts. This enables the government to avoid duplicating high start-up costs with another manufacturer.

MILESTONES FOR THE ACQUISITION CYCLE

Engines	
FY 1980 + FY 1981 Option	
Proposal received	March 11, 1980
Audit report recieved	April 30, 1980
Revised proposal received	July 11, 1980
Negotiations complete	Sept. 12, 1980
Contract distribution	Sept. 27, 1980
Exercise FY 1981 option	Nov. 14, 1980
FY 1982 Buy	
Proposal received	Jan. 31, 1981
Audit report received	March 27, 1981
Negotiations complete	May 29, 1981
Contract distribution	Oct. 15, 1981

Airframe	
FY 1980 Buy	
Proposal received	Aug. 29, 1980
Audit report received	Oct. 15, 1980
Negotiations complete	Nov. 27, 1980
Contract distribution	Dec. 22, 1980
FY 1981 + FY 1982 Option	
Proposal received	Aug. 12, 1981
Audit report received	Sept. 30, 1981
Negotiations complete	Nov. 6, 1981
Contract distribution	Dec. 2, 1981
Exercise option	Jan. 29, 1982

DEPARTMENT OF THE AIR FORCE
DETERMINATION AND FINDINGS

AUTHORITY TO NEGOTIATE AN INDIVIDUAL CONTRACT

Upon the basis of the following findings and determination, which I hereby make as agency head, the proposed contract described below may be negotiated without formal advertising pursuant to the authority of 10 U.S.C. 2304(a)(14).

FINDINGS

1) The Air Force Systems Command proposes to acquire by negotiation 60 General Electric F404/F1D2 Turbofan nonafterburning engines and associated engine containers, aeronautical support equipment, contractor support, training, initial spares and repair parts, and data at a total estimated price of $131,000,000. The F1D2 engine is deemed to be of a technical or special nature within the meaning of paragraph 3.214 of the defense acquisition regulation.

2) Acquisition by negotiation of the above described equipment is necessary because the equipment is so intricate and complicated that its manufacture requires a substantial initial investment and if manufactured by any supplier other than General Electric, Aircraft Engine Group would duplicate high starting costs already borne by the government. Starting costs such as preliminary engineering and development work and other manufacturing preparations by General Electric, Aircraft Engine Group, the developer and only producer of this equipment have been recovered by the contractor under other contracts with the government. Generally, such work would not be useful to or usable by any other supplier. Therefore, manufacture of this equipment by any other supplier will result in a duplication of these costs and would likely result in additional costs to the government by reason of such duplication.

3) Use of formal advertising for acquisition of the above described equipment is impracticable because such method may result in acquisition from another source which would require duplication of necessary initial investment to produce this equipment and would likely result in additional cost to the government by reason of such duplication.

DETERMINATION

The proposed contract is for equipment of a technical or specialized nature which requires a substantial initial investment to manufacture, and acquisition thereof by formal advertising would likely result in additional cost to the Government by reason of duplication of investment.

DEPARTMENT OF THE AIR FORCE DETERMINATION AND FINDINGS

AUTHORITY TO NEGOTIATE AN INDIVIDUAL CONTRACT

Upon the basis of the following findings and determination, which I hereby make as agency head, the proposed contract described below may be negotiated without formal advertising pursuant to the authority of 10 U.S.C. 2304(a)(14).

Findings

1) The Air Force Systems Command proposes to acquire by negotiation 20 Advanced Technology Aircraft and related initial spares and repair parts, aerospace support equipment, contractor support, training, and data at a total estimated price of $745,200,000. The Advanced Technology Aircraft is deemed to be of a technical or special nature within the meaning of paragraph 3.214 of the defense acquisition regulation.

2) Acquisition by negotiation of the above described equipment is necessary because the equipment is so intricate and complicated that its manufacture requires a substantial initial investment and if manufactured by any supplier other than Lockheed Corporation would duplicate high starting costs already borne by the government. Starting costs such as preliminary engineering and development work and other manufacturing preparations by Lockheed Corporation, the developer and only producer of this equipment, have been recovered by the contractor under other contracts with the government. Generally, such work would not be useful to or usable by any other supplier. Therefore, manufacture of this equipment by any other supplier will result in a duplication of these costs and would likely result in additional costs to the government by reason of such duplication.

3) Use of formal advertising for acquisition of the above described equipment is impracticable because such method may result in acquisition from another source which would require duplication of necessary initial investment to produce this equipment and would likely result in additional cost to the government by reason of such duplication.

Determination

The proposed contract is for equipment of a technical or specialized nature, which requires a substantial initial investment to manufacture, and acquisition thereof by formal advertising would likely result in additional cost to the government by reason of duplication of investment.

Appendix F

Pilot's Report on F-117A First Flight[54]

Take off time: 0605	Pilot: H. Farley	Date: June 18, 1981	
Landing time: 0619	Take off gross weight: 37,500 lb	Purpose First flight	Flight No.: 1
Total time: 0.3	Take off C.G. (gear down): 39.9%		A/C shop No.: 780

Configuration:
First flight—Fuel in forward and feed tanks

Summary

A successful first flight was accomplished achieving the planned objectives. The α and ß input to the flight control computer were turned on with satisfactory results. The maximum altitude achieved was 14,800 ft and the maximum speed was 254 KIAS (262 kn reported from the pace T-38). Directional stability was lower than expected requiring use of the ß input earlier than anticipated. Roll response was pleasant, but greater than expected. Pitch stability and response were excellent, closely matching our simulation. The flight was shortened due to a canopy unsafe light.

Discussion

Flight Description

Takeoff was accomplished on runway XX. Outside air temperature was 60°F and the winds were calm. The engines were heat soaked 4 min and 15 seconds prior to brake release. Acceleration was as expected reaching 140 KIAS at 3000 ft. Anticipated rotation was 185 KIAS. However, the airplane flew on its own at approximately 170–175 KIAS. Slight forward pressure was applied to hold a low angle of climb and accelerate. Shortly after lift off, the ß began to build up and rudder corrections were applied. The ß first went ~2 deg nose right then 3 deg nose left. When attempting to correct back to zero, the nose continued right to over 4 deg at which time the ß input to the flight control computer was turned on. This "stiffened" the directional axis significantly and the climb was continued at 250 KIAS. The pace T-38 reported 262 kn when I was reading 254. I reduced climb speed to an indicated 240–245 kn. A turn was commenced at ~8000 ft and 7 miles from the field. The 180 deg turn back toward the field was completed before reaching the planned altitude of 15,000 ft.

At this point, several events occurred requiring modification to the planned profile. The first thing that occurred was a canopy unsafe light. It went out shortly thereafter and cycled on and off several times throughout the flight. ECS surging also occurred at this time. The test conductor then reported higher than desirable temperatures in the nozzle area requiring a reduced thrust setting. Coupled with all of this were some loud "thumps" that I could not associate with a specific cockpit action. [It is now speculated that this noise was the blow-in doors opening or closing.] With this combination of problems, it was decided to fly the short profile. While intercepting and flying a right downwind entry, MMLE [modified maximum likelihood estimator]* control inputs were accomplished before and after turning on ARI [aileron rudder interconnect] and α input to the flight control computer.

A long straight in approach was flown starting out at 8 deg α, which resulted in over 200 KIAS. I felt this was too fast and decided to use 185 KIAS and accept the higher (~9 to 9.5 deg) α. Touchdown was smooth and the nosewheel remained in the air until intentionally lowered. The drag chute performed flawlessly and was retained to a full stop. Touchdown occurred 14 min after takeoff.

Directional Axis

The directional stability with the β probe off was not dissimilar to the NT-33 in configuration C. (Cn_{β} reduced by 50%). With the β probe on by itself the airplane felt slightly less stable than the simulator.

Roll Axis

Roll response was quicker than expected compared to the simulator. I had expected to select normal gain for the landing, however, the initial low gain setting was quite satisfactory for the mild maneuvering used on this flight.

Pitch Axis

It was difficult to tell the difference in pitch response to pulses before and after α probe turn on. It was highly damped and well behaved. This was

*MMLE inputs are standard pilot control inputs that should produce some calculated aircraft response outputs. These outputs can be used to calculate the aircraft's stability derivatives for comparison with predictions.

evident in the very pleasant speed and α control on final. I was able to promptly adjust airspeed from 200 KIAS to 185 KIAS and hold it to within 1 kn. The α did not wander as we have seen in the simulator.

Takeoff Trim Setting

The takeoff trim setting of one unit nose up (~2 deg TEU) was apparently too much resulting in a premature liftoff. I recommend a zero unit trim setting for the next takeoff.

Takeoff Acceleration

One line check was used during the takeoff run and that was at the 3000 ft point. The airspeed was 140 kn, which compares favorably with the acceleration prediction for an OAT [outside ambient temperature] of 60°F.

Boom Airspeed

The boom airspeed was used exclusively throughout the flight. A comparison between the pace and Senior Trend shows approximately 8 kn difference (Senior Trend reading lower) at 254 KIAS. Between 200 and 230 KIAS there was only a 2 kn difference and below 200 kn the pace and Senior Trend were reading the same.

Landing Gear Dynamics

The gear strut dynamics appear to be excellent. Touchdown was accomplished without any tendency to bounce and there was little or no impact transmitted into the airplane.

ECS

ECS [environmental control system] surging became evident near the top of the climb. It was not uncomfortable but was distinctly audible. I suspect it may have contributed to the intermittent canopy unsafe light due to the canopy seal inflating and deflating.

Canopy

The canopy unsafe light was my prime source of concern on the flight. We have to correct this prior to flight number 2.

"Thumps"

A loud thumping sound occurred approximately six times during the flight. I first thought it could be the ECS causing "oil canning." I also associated it with the landing gear down "slap" we experienced in our previous engine runs. It was suggested after the flight that this noise was the blow-in doors opening or closing and in retrospect, I suspect that is what it was. If that is the case, we should consider putting dampers on them.

Nozzle Temperatures

At the time of this writing, I am not completely informed on the detail of the high temperatures in the nozzle area. However, this appears to be the major problem as far as continuing the flight test program. We have to either correct the problem or decide we can tolerate higher temperatures.

Conclusions and Recommendations

1) The airplane was exceptionally clean for a first flight.
2) All major objectives were accomplished.
3) Directional stability is significantly lower than predicted.
4) Roll response is better than predicted.
5) Pitch axis and speed and α control was better than anticipated and was excellent.
6) Takeoff trim should be reduced to zero for the next flight.
7) Takeoff acceleration was right on the predicted value.
8) Boom airspeed is accurate up to 200 kn, then has an increasing error (reads low) of up to 8 kn at 250 KIAS.
9) Landing gear dynamics are excellent.
10) The canopy unsafe light must be corrected prior to the next flight.
11) The nozzle temperature problem requires correction before the next flight.

[signed] Hal Farley

Appendix G

Summary of Costs and Contracts

This appendix summarizes the major contracts awarded and costs incurred during the Have Blue and Senior Trend programs. For Have Blue, all phases of the program are addressed, from preliminary studies through the completion of flight testing. In the case of the Senior Trend program, discussion is limited to those costs associated with getting the system into operation with approximately the capability initially envisioned, including research and development, procurement, and post-initial operational capability development through approximately 1990, but not operations and support or recent upgrades.

All costs in this appendix are expressed in then-year dollars. Unless otherwise referenced, the information is drawn from a Lockheed document, "F-117A Cost Performance and Contracts History," dated April 19, 1989, which includes information on the Have Blue program as well as on the F-117A. A cover note written by Col Mike Marks, SAF/AQL,* dated May 3, 1989, states: "This is an important historical document.... It represents only the contractors view but is still as close to events as I remember them."[128]

Have Blue

Initial 1974 DARPA study contracts were approximately $100,000 each to McDonnell Douglas and Northrop. Contract type was firm fixed price (FFP). Lockheed entered this phase late and at no cost to the government.[188, 234]

Phase I contracts to Lockheed and Northrop were FFP, approximately $1.5 million each for preliminary design work plus the construction of large-scale RCS models for a competitive evaluation at RATSCAT (see Table G1 for the exact value of Lockheed's contract). The only change ($0.187 million to Lockheed) was for the construction of a new model support system (pylon) to allow accurate measurements at lower RCS values than previously required.

* SAF/AQL is the office symbol for Assistant Secretary of the Air Force (Acquisition), Directorate of Special and Electronic Programs. This office has the responsibilities formerly held by AF/RDPJ, the original Air Force low observables office in the Pentagon. It was also referred to for a time as RDQ-LO. Col Marks was the director of SAF/AQL in 1989 and was previously a member of the RDPJ staff during the early stages of the Senior Trend program.

The phase II contract to Lockheed was originally negotiated as cost plus fixed fee (CPFF) with a ceiling of $19.23 million. It was understood that Lockheed would also contribute approximately $10 million. The phase II contract covered the design, development and fabrication of the two Have Blue aircraft and the first 10 months of flight testing, referred to as "basic" flight test. The cost of phase II had been the subject of extensive negotiations from March through June 1976. The phase II "ceiling" of $19.23 million was established on June 4.

Authorized scope changes in phase II totaled $6.4 million. These included some items initially negotiated as government furnished aerospace equipment, subsequently furnished by Lockheed; changing the flight test location from Palmdale where it had been originally planned, to the classified test site; additional technology studies; and expanded wind tunnel, RCS, and flight control activities. Upon completion of the "basic" flight testing in September 1978, the phase II contract was renegotiated and closed as firm fixed price at $27.5 million, which included the $6.4 million scope changes and a $1.8 million overrun. Lockheed shared costs of $10.1 million beyond this final price, in the form of a profit write-off each month.[188]

Follow-on testing starting October 1, 1978, was funded under a new contract. Follow on testing continued through July 1979 when the second Have Blue aircraft was lost. The Lockheed contracts and costs are summarized in Table G1.

Only two major subcontracts were involved in Lockheed's Have Blue effort: Lear Siegler for the modified F-16 flight control system ($1,081,000), and General Electric for J-85 engine modifications and testing ($200,000). Lockheed rated both subcontractors' cost, schedule, and technical performance as "excellent."

In addition to the $32.59 million in contracts to Lockheed, government costs included approximately $10 million for government furnished aerospace equipment,, test support, program office, and others. Of these costs, DARPA funded $23.2 million, and the Air Force provided $19.4 million.

Table G1 Lockheed contract costs in the Have Blue program[88, 126, 128]

	Phase I	Phase II	Follow on test	Total
Contract type	FFP	CPFF	CPFF	
Originally negotiated price	1.425	19.230	5.609	26.264
Authorized scope changes	0.187	6.413	–1.361	5.239
Overrun	—	1.835	–0.748	1.087
Final contract price	1.612	27.478	3.500	32.590
Lockheed cost sharing	0.240	10.100	—	10.340
Total	1.852	37.578	3.500	42.930

Adding the $1.7 million in early contracts to other companies (the preliminary study contracts plus Northrop's phase I contract), plus the $10.3 million Lockheed cost sharing, gives a total cost for the entire program of $54.6 million then-year dollars. This compares favorably to other technology demonstrator aircraft programs.[10]

SENIOR TREND

A one-year concept definition contract for $11.1 million was awarded to Lockheed on October 10, 1977. The final price of this contract was apparently $10.1 million. A possible cause of the underrun may be that the choice of concepts was narrowed from two (ATA "A" and ATA "B") to one during the summer of 1978, several months before the completion of the contract period.[115]

As the study phase reached completion, full-scale development was initiated. At the time of full-scale development go-ahead in November 1978, Lockheed ADP estimated the total full-scale development costs at $504.4 million including program office, test support, government furnished equipment, and others, as well as the primary contract with Lockheed.[195] The original negotiated value of the full-scale development contract to Lockheed was $340.1 million. This cost plus fixed fee (CPFF) contract covered five full-scale development aircraft, spares, support, and flight test.

Engines were acquired as government furnished equipment. Thirteen F404 engines were procured for the full-scale development program; they were developed under a separate CPFF contract with the Aircraft Engine Division of the General Electric Company.[66, 67] Engines cost approximately $2 million each during the period in which they were procured for the full-scale development and early production F-117s.[30, 43]

Costs soon increased, due to a combination of economic factors and development difficulties.[90] By the time limited initial operational capability was achieved in October 1983, the final full-scale development contract price was $763 million. Negotiated scope changes (engineering change proposals, spares,and so on) totaled $135.7 million. The remaining cost growth (the difference between $763.1 and $475.8 million) is attributed by Lockheed to a number of causes including "external factors" and design and manufacturing problems. It also includes flight test problems. The results were increases of $100.5 million (or 21% of the negotiated scope) attributed to external factors (such as unforeseen changes in inflation, long material lead times, etc.); $103.5 million (or 22%) to design and manufacturing problems; and $83.3 million (18%) to flight test problems and associated delays. Thus, the total cost growth was approximately 60% relative to the final negotiated scope of the full-scale development contract. Nevertheless, the program costs compare favorably with most other recent military aircraft development programs. The concept

Table G2 Concept definition and full-scale development cost summary

	Concept definition	Full-scale development	Total (FY $M)
Contract type	CPFF	CPFF	
Original negotiated value	11.1	340.1	351.2
Negotiated scope changes ("set-asides," ECPs, spares, etc.)	—	135.7	135.7
Subtotal (negotiated scope)	11.1	475.8	486.9
Cost growth			
External factors	—	100.5	100.5
Design manufacturing problems	–1.0	103.5	102.5
Flight test problems	—	83.3	83.3
Final price	10.2	763.1	773.2

definition and full-scale development contract costs are summarized in Table G2. (Final *total* costs including government test support, goverment furnished equipment, and others are probably not available.)

Production Lots 2 through 6 (the first five production lots, because the full-scale development aircraft were considered Lot 1) were fixed price incentive (FPI) contracts with ceilings that reflected a progressively lower unit cost. Lockheed had initially wanted cost-reimbursement–type contracts for early production, but Gen Alton Slay (commander of the Air Force Systems Command) reportedly insisted on a fixed price. Lockheed's performance on those production contracts that were FPI is shown in Table G3. The last five lots were firm fixed price (FFP). Each production contract included spare parts, support equipment and authorized engineering change proposals plus deliverable F-117As.

New contracts added during the program included one for full-capability development and one for a weapon system trainer. Both were FPI-type contracts. At around Lot 4, the Air Force authorized a $20 million producibility improvement program "based on a Lockheed commitment to show at least a

Table G3 Lockheed's performance on FPI production contracts

Lot no.	Overrun/(underrun)
2	37.0
3	22.0
4	(13.0)
5	(3.0)
6	(18.0)
Total	25.0

two-to-one payoff in reduced production costs. It is generally agreed that the company bettered that objective."

An overall summary of Senior Trend contracts is presented in Table G4.

Table G4 Total program costs

Senior Trend Program Contracts (TY$M)	Contract Type	Negotiated Scope	Final Price	Overrun
Concept definition	CPFF	11.1	10.1	–1.0
Full-scale development (FSD)	CPFF	475.8	763.1	287.3
Full capability development (FCD)	FPI	617.6	611.4	–6.2
Weapon system trainer	FPI	88.0	91.0	3.0
Other development costs thru ~1990			~524.0	
Total development cost thru ~1990			1999.6	
Production lots 2–6	FPI	XX[a]	XX	25.0
Production lots 7–11	FFP	XX	XX (fixed price)	
Producibility improvement program		~20.0	~20.0	~0.0
Total flyaway cost			2514.6	
Total procurement cost			4265.3	
MILCON (military construction)			259.4	
Total program acquisition cost			6560.3	
Unit flyaway cost (incl. engines)			42.6	
Unit program cost			111.2	

[a]XX signifies data not available.

Appendix H

Have Blue and Senior Trend Time Table

The authors do not claim that this chronology is exhaustive. The intent is to provide perspective by providing a chronological listing of the most significant events focusing most heavily on the period up to and including limited initial operational capability (LIOC) of the F-117A in 1983.

Reference citations appear throughout the main text and therefore are not repeated here. *Some* additional information and events from the more recent years have also been included in this chronology. Reference 232 is the source for these additional items.

The tailpipe/platypus, air data probes, inlet grids, RAM coating materials and processes, fuel system, and various airframe items were the subject of continual developmental test and evaluation from early full-scale development through LIOC, and of specific improvement programs after LIOC, some of which continue. Some, but not all, of these modification and upgrade programs also are listed here.

19th Century Maxwell's equations describing electromagnetic fields and waves are formulated. Some exact solutions found for electromagnetic scattering from simple shapes. Sommerfeld determines exact solution for two-dimensional diffraction from a wedge.

1903 First radar experiments.

1930 First aircraft detection by radar.

1930s Practical radar systems developed for military applications.

1935 Radar cross section (RCS) reduction for aircraft first suggested by Sir Robert Watson-Watt, British radar pioneer.

Late 1930s–1950s Geometric and physical optics theories developed for the approximate calculation of RCS. (In the case of geometric optics, the theory was much older, but the application to RCS calculation was new.)

1940 onward	Rapid advances in radar technology and applications during World War II and continuing afterward.
1943	Lockheed Advanced Development Projects (the Skunk Works) established to design and build the XP-80 jet fighter.
1944–1960	First attempts to reduce RCS of aircraft, primarily by parasite treatments (add-on RAM). Various types of RAM developed for aircraft and nonaircraft uses.
1950s	Geometric and physical theories of diffraction developed.
	First RCS measurement ranges constructed.
1962	*Method of Edge Waves in the Physical Theory of Diffraction* published in Soviet Union by P. I. Ufimtsev.
1960s	Aircraft began to be designed from the ground up for low RCS, incorporating shaping as well as RAM. Details also worked extensively (inlets, canopies, apertures, and others). Effectiveness of shaping was limited by lack of theoretical understanding.
1965–1972	Vietnam War illustrated growing impact of radar threats to aircraft.
1966–1970	Northrop developed "first-generation" RCS prediction computer programs under Air Force Avionics Laboratory contracts. Notable capabilities included modeling of inlet cavities, radar absorbent materials, reflection and diffraction calculations.
1971	*Method of Edge Waves...* translated by U.S. Air Force Systems Command's Foreign Technology Division. (However, Northrop was already developing RCS prediction methods that used various theories, including Ufimtsev's.)
Early 1970s	Various researchers determined an explicit solution for the physical optics scattering from a polygonal flat plate. Northrop developed improved RCS prediction code (GENSCAT). Successfully calculated the RCS of F-4 aircraft. Published documentation and supporting theoretical papers under Air Force contract, "Calculation of Radar Cross Section." Teledyne Ryan developed low-RCS drone concepts using aggressive shaping of the airframe and some analytical techniques to design edge treatments.

1973 Yom Kippur War dramatically illustrated vulnerability of latest Western tactical aircraft to Soviet radar-guided missiles and guns. Aircraft survivability recognized as critical issue to balance of power between NATO and Warsaw Pact.

1974 DARPA initiated studies to identify RCS thresholds for avoiding detection and means of achieving such thresholds. Resulted in $100,000 contracts to Northrop and McDonnell Douglas.

1975

Jan./Feb. Lockheed learned of DARPA program, obtained permission to participate on basis of prior RCS reduction experience on A-12 and SR-71 manned reconnaisance aircraft and D-21 drone.

Spring Based on input from the study participants, DARPA established RCS goals for the program.

Lockheed designers identified faceted shapes as a way to achieve extremely low RCS and developed a preliminary faceted configuration (the "Hopeless Diamond").

Lockheed developed ECHO I computer program based on the physical optics solution for a flat panel. Using ECHO, designers refined the "Hopeless Diamond" into a viable aircraft design.

Concurrently, a "Hopeless Diamond" model was built and tested, providing some validation of initial predictions but also revealing serious shortcomings in the ability of existing facilities to measure very low RCS levels.

Lockheed and Northrop presented faceted aircraft concepts to DARPA and predicted that RCS goals would be met.

DARPA urged McDonnell Douglas to team with Teledyne Ryan to benefit from Ryan's low RCS drone experience.

Summer DARPA formulated XST program strategy: phase I of competitive ground RCS tests; winner to be funded for phase II development and flight test of manned demonstrator aircraft.

Air Force Chief of Staff promised U.S. Air Force support for demonstration program.

Contractors stated willingness to share costs if selected for phase II.

August DARPA requested proposals from the three current participants.

October	Proposals reviewed. Security increased based on DARPA and the Office of Defense Director of Defense, Research, and Engineering discussions.
Oct. 1–3	Radar Camouflage Symposium was held at Wright–Patterson Air Force Base.
Nov. 1	XST phase I contracts of $1.5 million each were awarded to Lockheed and Northrop. Both contracts contained phase II option.
Nov.–Dec.	Contractors proceeded with additional design work and subscale model testing in preparation for phase II selection. XST participants identified need for improved model support system (pylon) at RATSCAT; joint design effort followed, with fabrication by Lockheed ADP.
1976	
January	New very low signature support system delivered to RATSCAT.
March	Competitive "Pole-Off" between Lockheed and Northrop, and evaluation of proposals that included additional design data (aero, performance, and others) and program plans.
April	"Pole-Off" results led to increased high-level Air Force and the Office of the Secretary of Defense interest in the program. Program renamed Have Blue and placed under Air Force management as a special access program to protect technological breakthrough.
April 26	Lockheed selected and authorized to proceed to phase II. DARPA encouraged Northrop XST team to remain ready for as-yet undetermined future program(s).
May–Sept.	High-speed, low-speed and propulsion (inlet) wind tunnel testing in support of Have Blue demonstrator final design effort.
May–Dec.	Additional RCS testing at Grey Butte and RATSCAT.
June 4	Phase II contract negotiations completed; ceiling set at $19.23 million.
July	Fabrication began for Have Blue No. 1.
August	Air Force headquarters initiated preliminary requirements studies to apply the breakthrough to one or more operational weapons systems.

September	Lockheed requested additional funds for Have Blue scope changes.
October	Have Blue flight control laws defined.
1976–1978	Lockheed RCS prediction research continued, capabilities added to ECHO computer program: edge diffraction calculations based on Ufimtsev's *Physical Theory of Diffraction*, together with some ability to include the effects of nonmetallic materials. Northrop continued RCS theory/computational work and design studies, initially unfunded and later with DARPA funding.
1977	
January	First full-scale RCS tests of final Have Blue design.
March	Have Blue flight control system tests.
April/May	Air Force headquarters asked Air Force Systems Command for survey of technology programs addressing missionization issues for a very low observable aircraft (targeting, avionics emissions, and others).
May	Assembly of Have Blue No. 1 began.
June	Project office established within Air Force headquarters at the Pentagon (AF/RDPJ) to conduct planning and concept exploration for low observable weapon systems. B-1A program canceled. Strategic Air Comand sought potential replacement.
July	Lockheed costs tracking above available funding for Have Blue. Air Force requested assurance that Lockheed would not seek reimbursement for costs above ceiling.
August	Lockheed reaffirmed willingness to share costs of $10.4 million, plus company-funded development, facilities and equipment bringing total Lockheed contribution to around $14 million. Aerospace machinist's strike began. Lockheed Skunk Works managers worked to complete Have Blue No. 1.
Aug.–Sept.	Structural ground tests conducted on Have Blue No. 1.

Oct. 10	One-year Advanced Technology Aircraft (ATA) concept exploration contract awarded to Lockheed. Two sets of possible requirements: "A" model (5000 lb payload × 400 n miles mission radius) and "B" model (10,000 lb payload × 1000 n miles).
Nov. 4	First engine runs performed in Have Blue No. 1 at Lockheed Burbank.
Nov. 16	Have Blue No. 1 transported to test site by C-5 (originally scheduled for October but delayed by strike).
Dec. 1	Have Blue No. 1 first flight.
1978	DARPA initiated Tacit Blue program to demonstrate concept and technologies for a low observable reconnaisance aircraft. Northrop was prime contractor. Wind tunnel testing began in support of ATA design studies.
Winter/Spring	Risks evident with ATA "B" model. Payload reduced due to difficulty satisfying original requirements.
May 4	Have Blue No. 1 lost in landing accident. Thirty-six test sorties completed.
Summer	Decision by Air Force to proceed with development of smaller "A" model ATA. Preparations made to secure funding and process necessary program documentation for full-scale development start.
July 20	Have Blue No. 2 first flight.
Sept. 30	Have Blue "Basic" flight test completed, Follow-on testing began under new contract.
November	Senior Trend (ATA) full-scale development go-ahead and prime contract awarded to Lockheed (separate contract negotiated with General Electric for engines). First flight planned for July 1980, IOC for March 1982. Planned production quantity: 20 aircraft.
1979	
Early 1979	Management of Senior Trend and other low observables programs transferred from AF/RDPJ to the Aeronautical Systems Division, Wright–Patterson Air Force Base.

	RDPJ continued to perform planning and budgeting-type activities. Construction began of wooden mockup; wind tunnel testing of air data probes and inlets (including inlet icing tests); design of maintenance vans and manuals.
April	First RATSCAT entry for Senior Trend program (6/10 scale model).
July 11	Have Blue No. 2 lost following exhaust failure. Fifty-two sorties completed.
August	Various technical difficulties and challenges led to projected Senior Trend funding shortfalls.
October	Senior Trend full configuration wind tunnel test.
Oct. 15	4450th Tactical Group (which would operate the Senior Trend aircraft) formed under command of Col Robert A. Jackson.
December	Long-lead procurement authorized for first production lot. Structures fabrication began on first full-scale development aircraft. Advanced logistic support activities initiated. Official end of Have Blue program.
1980	Full authorization for first production lot. Air Force began selecting pilots for F-117A squadrons. Increased production run contemplated. Engine and exhaust system ground tests began. Lockheed began working to develop spray-on RAM capability. Lockheed ADP manpower shortages affected program schedule.
March–April	Fabrication of static test airframe began.
July	Originally planned first flight date.
Dec. 22	Tailpipe failure during engine/exhaust system ground testing.
1981	Efforts undertaken to resolve air data probe, exhaust system problems. Northrop won advanced technology bomber contract.

March	First full-scale development aircraft completed (Aircraft 780).
March–May	Long-lead planning to protect transition into increased production.
April 27–28	Aircraft 780 First Flight Readiness Review.
April 28	Tailpipe fire in Aircraft 780 during engine run.
June 18	F-117A first flight.
July	Air Force pressured Lockheed ADP to address manpower shortage.
Aug. 6	Tenth flight. Aircraft 780 grounded for scheduled maintenance, inspections, and larger tail fins.
August	Col Richard Scofield replaced Col Don Merkl as Senior Trend program manager.
Sept. 24	First flight, Aircraft 781.
Summer/Fall	Tentative approval for increased production run (89 aircraft). Final approval contingent upon successful RCS, avionics tests. Corresponding additions to program content including increased Air Force logistics support, weapon system trainer, automated mission planning.
Dec. 18	First flight, Aircraft 782.

1982

Jan. 23	First RCS test sortie (Aircraft 781. Pilot: Lt Col Roger Moseley).
February	Sherman Mullin became Lockheed ADP's F-117A program manager. Northrop's demonstrator aircraft for Tacit Blue began flight test.
March	Originally planned Senior Trend IOC date.
March 22	First F-117A night flight.
March–April	Reduced production procurement (57) appeared imminent.
April 10	First flight, Aircraft 784.

April 20	First production F-117A (Aircraft 785) crashes on first takeoff.
April 26	Air Force decided to reduce production quantity to 57 aircraft.
May 17	Col James S. Allen assumed command of 4450th Tactical Group. Move from test site to Tonopah initiated.
July 7	First Flight, Aircraft 783. All five full-scale development aircraft now flying.
Aug. 23	Aircraft 787 accepted by Air Force (delivered to the test force).
Sept. 2	4450th Tactical Group took delivery of its first F-117A, Aircraft 786.
Nov.–Dec.	Various DoD program budget decisions questioned Senior Trend funding.
1983	4450th Tactical Group completed move to Tonopah. Continued development testing, particularly on the air data probes, IRADS bay and apertures, and exhaust system.
Jan.–Sept.	Tactical Air Command aircraft modified to LIOC configuration.
June	All F-117As grounded for inspection of tailpipe support posts.
September	Configuration update 1 started. (Configuration updates incorporated various modifications and fixes into the F-117A fleet.)
Oct. 28	Limited IOC (LIOC) achieved.
1984	
April 1	Weapon system computational subsystem upgrade authorized.
June or July	First operational readiness inspection successfully passed.
July 1	Full-scale development completed. Full-capability development continued under a new contract.
Late 1984	"Quick-look" Navy evaluation of F-117A. Conclusion was that F-117 was not suitable for carrier operations.
1985	Tests of experimental F-117A radar antennas. Tacit Blue program ended.

Early January	Fabrication began of Aircraft 777 (fatigue test airframe).
Jan. 21	Aircraft 779 (static test airframe) stressed to failure.
July	Configuration update 2 started.
August	First automated RAM spray on major airframe component (wing).
Sept. 25	Tailfin failed in flight on Aircraft 781. Eventually led to design of new thermoplastic composite fin.
1986	Offensive Capability Improvement Program phase II started ("four-dimensional" flight management system, improved cockpit). GBU-27 precision glide bomb drop tests from F-117As.
Jan. 8	Configuration update 1 completed.
June	Configuration update 3 started.
June 11	First production aircraft with sprayed-on RAM.
July 11	First F-117A operational loss. Pilot Maj Ross Mulhare killed.
Aug. 8	Composite thermoplastic fins authorized.
Sept. 24	IRADS "Block -1" modification approved.
Oct. 2	First flight, weapons system computational subsystem upgrade.
Oct. 28	Configuration update 2 completed.
1987	
Feb. 20	Testing of Aircraft 777 (fatigue) began.
March	First Tactical Air Command aircraft received IRADS Block -1 modification.
April	Flight No. 10,000 completed.
July	15,000 total flight hours accumulated.
August	Configuration update 3+ started.

Sept. 24	First fatigue life completed, Aircraft 777.
November	First production delivered with weapons system computational subsystem upgrade. Configuration update 4 started.
1988	GBU-27 precision glide bomb certified with F-117A.
Jan. 15	Configuration update 3 completed.
Jan. 21	First weapons system computational subsystem upgrade retrofit.
May 3	Configuration update 3+ completed.
September	Last IRADS Block -1 modification completed.
Sept. 8	Fatigue test completed at 20,500 equivalent flight hours.
Nov. 10	Program officially acknowledged.
Nov.–Dec.	Composite fin fatigue and acoustic test articles delivered to Lockheed Rye Canyon research facility for test.
Dec. 1	First test flight, Offensive Capability Improvement Program phase II modifications.
1989	
May	Fatigue test aircraft (Aircraft 777) inspection completed.
July 18	First flight of thermoplastic tail fin.
Oct. 5	4450th Tactical Group renamed 37th Tactical Fighter Wing.
Oct. 12	Production Termination Program received go-ahead.
Dec. 21	First combat use of F-117As, in Operation Just Cause (Panama).
Late 1980s	F-117A achieved substantially improved reliability and maintainability performance relative to initial levels, for example, maintenance man-hours per flight hour reduced to approximately 45.
1990	
April 21	First public showing of F-117A.

April–Nov.	Installation of first Offensive Capability Improvement Program phase II modification kit to production aircraft.
May	Configuration update 5 started.
June 18	First production retrofit of composite fin.
July 12	Last (number 59) production F-117A delivered.
August	"F3" IRADS modification approved.
Oct. 1	Production Termination Program tooling transfer began.
Oct. 25	Configuration update 4 completed.
1991	
Jan. 17–Feb. 28	Operation Desert Storm. Forty F-117As flew approximately 1270 combat sorties against targets in Iraq with no loss or damage due to enemy action.
July 17	Aircraft 781 transferred to the Air Force Museum at Wright–Patterson Air Force Base.
Nov. 1	37th Tactical Fighter Wing renamed 37th Fighter Wing.
1992	
Jan. 21	Last weapons system computational subsystem upgrade retrofitted.
Feb. 28	Production Termination Program tooling storage completed.
May 16	Aircraft 780 mounted for display at Nellis Air Force Base.
May–July	37th Fighter Wing moved from Tonopah to Holloman Air Force Base, New Mexico, and renamed 49th Fighter Wing.
June 16	Last production (thermoplastic composite) fin delivered (but retrofit to Saudi Arabia deployed aircraft delayed until 1994).
July 31	Production Termination Program completed.
Aug. 14	First flight IRADS F3 turret.
1993	
Feb. 12	IRADS F3 flight test completed.

July	50,000th flight achieved.
August	80,000 total flight hours accumulated.
Nov. 25	Aircraft 837 joined test force for FOT&E.
1994	
Aug. 11	Last composite fin retrofit completed.
September	MIL-STD-1760 modification program initiated.
Sept. 7	Aircraft 777 (formerly fatigue test) conversion to pole model authorized.
Oct. 1	First production IRADS F3 retrofitted.
October	Configuration update 5+ started.
1995	
March 8	Configuration update 5 completed. Last Offensive Capability Improvement Program phase II modification completed.
Aug. 30	Conversion of Aircraft 777 to pole model completed.
1996	
October	IRADS F3 retrofits completed.
November	Configuration update 6 started.
December	Configuration update 5+ completed.
Ongoing as of July 1997	Tailpipe/ platypus improvements; inlet grid improvement; training and mission data planning system improvements; weapon system trainer updates; MIL-STD 1760 modification; low observables communications; configuration update 6; various additional improvement programs (MIPS, RNIP, CATE, DIR)

References

Note: Many of the references originally carried classification markings; however, all information used in this book has been approved for release as *unclassified.*

[1]Abrams, R., (Lockheed ADP), Memorandum to Norm Nelson, Alan Brown and Keith Beswick (all Lockheed ADP), "Daily Status Report June 12 through June 14, 1981," June 15, 1981.

[2]Abrams, R., (Lockheed ADP), Memorandum to Norm Nelson, Alan Brown and Keith Beswick (all Lockheed ADP), "Daily Status Report for June 15, 1981," June 16, 1981.

[3]Abrams, R., (Lockheed ADP), Memorandum to Norm Nelson, Alan Brown and Keith Beswick (all Lockheed ADP), "Daily Status Report for June 16, 1981," June 17, 1981.

[4]Abrams, R., (Lockheed ADP), Memorandum to Norm Nelson, Alan Brown and Keith Beswick (all Lockheed ADP), "Daily Status Report for June 18, 1981," June 19, 1981.

[5]"Air Force Unveils TACIT BLUE Stealth Aircraft," Air Force Link Internet Release 01-04-96, April 30, 1996.

[6]*Air-to-Surface Technology Integration and Evaluation Study, Vol. 4—Stealth*, Rept. No. FZA-479-3/4, General Dynamics, Fort Worth, TX, Jan. 15, 1977.

[7]Allen, J. S., Brig Gen, U.S. Air Force (Ret.), Comments provided to the authors, March 8, 1996.

[8]Allen, J. S., Col, U.S. Air Force, *4450 Tactical Group Commander's Declaration of IOC*, Oct. 28, 1983.

[9]Anderson, S., Col, U.S. Air Force (Senior Trend JTF), Flight Report: *A/C 780 Flight No. 7*, July 28, 1981.

[10]Anderson, S., Col, U.S. Air Force, Flight Report: *A/C 780 Flight No. 8*, July 30, 1981.

[11]Angelucci and Bowers, *The American Fighter*, Orion Books, New York, 1985.

[12]Aronstein, D. C., and Piccirillo, A. C., *Comments and Observations on Flight Demonstrator Programs*, ANSER, Inc., Arlington, VA, Oct. 1994.

[13]Aronstein, D. C., and Piccirillo, A. C., *The Lightweight Fighter Program: A Successful Approach to Fighter Technology Transition*, ANSER, Inc., Arlington, VA, March 1995.

[14]Aronstein, D. C., "The Development and Application of Aircraft Radar Cross Section Prediction Methodology," AIAA/SAE World Aviation Congress (Paper No. 965539), Los Angeles, Oct. 1996.

[15]*Aviation Week and Space Technology*, Aug. 2, 1976, p. 11.

[16]*Aviation Week and Space Technology*, Feb. 14, 1977, p. 9.

[17]*Aviation Week and Space Technology*, June 20, 1977, p. 11.

[18]*Aviation Week and Space Technology*, Aug. 11, 1980, p. 17.

[19]*Aviation Week and Space Technology*, Sept. 1, 1980, p. 41.

[20]*Aviation Week and Space Technology*, Sept. 8, 1980, pp. 15, 31.

[21]*Aviation Week and Space Technology*, Oct. 12, 1981, p. 17.

[22]Baber, J. W., Lt Col, U.S. Air Force (Senior Trend SPO), Memorandum to Col Ken Staten (RDPJ), "What-If Exercise No. 3060," Nov. 12, 1980.

[23]Baber, J. W., Lt Col, U.S. Air Force (Senior Trend SPO), Memorandum to Lt Col Mike Marks (RDPJ), "Senior Trend Funding," undated but in response to inquiry of April 22, 1981.

[24]Baber, J. W., Lt Col, U.S. Air Force (Senior Trend SPO), Memorandum to Mr. Passon (Lockheed ADP), info to Lt Cols Mike Marks and John Madia (RDPJ), Aug. 23, 1982.

[25]Baber, J. W., Col, U.S. Air Force (Ret), Comments provided to the authors, July 15, 1996.

[26]Bahret, W. F., AFAL/WRP, "Introduction to Radar Camouflage," *Proceedings of the 1975 Radar Camouflage Symposium*, Air Force Avionics Laboratory Technical Rept. AFAL-TR-75-100, Dec. 1975.

[27]Battershell, A. L., "Technology Approach: DoD Versus Boeing (A Comparative Study)," *Acquisition Review Quarterly*, Summer 1995, pp. 214–227.

[28]Bertelson, R. K., (General Electric), Letter to Col Don Merkl (Senior Trend SPO), "Follow-on F404/F1D2 Production," March 4, 1981.

[29]Bird, J., "Now the Secret is Out," *Air Force Times*, May 13, 1996.

[30]Bond, R. M., Lt Gen, U.S. Air Force (AFSC/CV), Memorandum to U.S. Air Force Headquarters/RD, Subject: F404 cost growth, Jan. 13, 1982, with handwritten corrections.

[31]Bowlds, T. F., Col, U.S. Air Force, Rome Laboratory commander, Letter to Albert C. Piccirillo, May 28, 1996.

[32]Brown, A., (Lockheed ADP), *Engine/Exhaust System Overheat Problem*, April 28, 1981.

[33]Brown, A., "Stealth Technology Leading to the Lockheed F-117A Stealth Fighter," AIAA/SAE World Aviation Congress, Los Angeles, Oct. 24, 1996.

[34]Brown, A., Comments provided to the authors, Oct. 23, 1996.

[35]Brucker, L., United States Air Force Aeronautical Systems Center, Wright–Patterson AFB, OH, Comments provided to the authors, March–April 1996.

[36]Burrows, W. E., "How the Skunk Works Works," *Air and Space*, April/May 1994, pp. 30–41.

[37]Casper, J., Lt Col, U.S. Air Force, Memorandum to Lt Col Bill Trembour, "F-117 vs. SU AWACS," DACOM No. 07-080, July 21, 1982.

[38]Chun, C. K. S., *The Lockheed F-117A*, RAND, Santa Monica, CA, 1991.

[39]Coleman, J. R., Heath, H. C., and Oshiro, F. K., *Radar Cross Section Calculations, Part I*, Rept. No. NOR-76-214.1, Northrop Corp., Hawthorne, CA, Dec. 1976.

[40]Cornelisse, D. G., chief, ASC History Office, *The Legacy of Stealth: The Development of Low Observable Aircraft at the Aeronautical Systems Center,* Wright–Patterson AFB, OH, (DRAFT), Oct. 4, 1996.

[41]Covault, C., "Advanced Bomber, Missile in Definition," *Aviation Week and Space Technology*, Jan. 29, 1979, pp. 113–121.

[42]Crispin, J. W., Jr., Goodrich, R. F., and Siegel, K. M., *A Theoretical Method for the Calculation of Radar Cross Section of Aircraft and Missiles*, Univ. of Michigan Radiation Laboratory, Ann Arbor, MI, Rept. 2591-1-H, July 1959.

[43]Department of the Air Force, *Determination and Findings: Authority to Negotiate on Individual Contract (Engines)*, Nov. 28, 1978.

[44]Department of the Air Force, *Determination and Findings: Authority to Negotiate on Individual Contract (Engines)*, Sept. 30, 1980.

[45]Department of the Air Force, *Determination and Findings: Authority to Negotiate on Individual Contract (Engines)*, June 15, 1981.

[46]Department of the Air Force, *Determination and Findings: Authority to Negotiate on Individual Contract (Engines)*, Nov. 19, 1982.

[47]Directorate of Special Management, SM-ALC/QL, *F-117A Budget Defense Document*, July 1991.

[48]Donald, D., and Lake, J., Ed., *Encyclopedia of World Military Aircraft*, Vol. 2, Aerospace Publishing Ltd, London, 1994.

[49]Dorr, R. F., *Lockheed F-117 Nighthawk*, Aerospace Publishing Ltd, London, 1995.

[50]Dyson, K., Lt Col, U.S. Air Force (Ret), Comments provided to the authors, Aug. 1996.

[51]Ethell, J., *Mustang: A Documentary History of the P-51*, Jane's, London, 1980.

[52]Evans, G. R., *Radar Absorbing Materials (RAM): An Annotated Bibliography of Report Literature*, Lockheed Missiles and Space Co., Palo Alto, CA, March 1964.

[53]Fagnant, J., Maj U.S. Air Force (SAF/AQL), F-117A Program Element Monitor (PEM), Comments provided to the authors, Dec. 1995.

[54]Farley, H. (Lockheed ADP), Flight Rept.: *A/C 780 Flight No. 1*, June 18, 1981.

[55]Farley, H., Flight Rept.: *A/C 780 Flight No. 2*, July 1, 1981.

[56]Farley, H., Flight Rept.: *A/C 780 Flight No. 3*, July 9, 1981.

[57]Farley, H., Flight Rept.: *A/C 780 Flight No. 4*, July 15, 1981.

[58]Farley, H., Flight Rept.: *A/C 780 Flight No. 5*, July 23, 1981.

[59]Farley, H., Flight Rept.: *A/C 780 Flight No. 6*, July 25, 1981.

[60]Ferguson, D., (Lockheed ADP), Flight Rept.: *A/C 780 Flight No. 9*, Aug. 1, 1981.

[61]Ferguson, D., Flight Rept.: *A/C 780 Flight No. 10*, Aug. 6, 1981.

[62]Ferry, Capt, U.S. Air Force (Senior Trend SPO), Memorandum to Lt Col Mike Marks (RDPJ), "Changes to FY84 from 19 Oct 81 Submission," DACOM No. 006178, March 15, 1982.

[63]Fulghum, D. A., "Secret Flights in 1980s Tested Stealth Recon," *Aviation Week and Space Technology*, May 6, 1996.

[64]General Accounting Office, *History of the F-117 Program*, Draft, GAO/C-NSIAD-90-XXX, Jan. 1990.

[65]General Accounting Office, *Uncertainties Persist in the Advanced Cruise Missile Program*, GAO/NSIAD-91-35, Nov. 1990.

[66]General Electric, *Workscope for the F404/F1D2 Engine Program*, Oct. 1978.

[67]General Electric/U.S. Air Force, *Pre-Letter Contract Negotiations*, Oct. 5, 1978.

[68]Giangreco, D.M., *Stealth Fighter Pilot*, Motorbooks International, Osceola, WI, 1993.

[69]Gordon, W. B., "Far Field Approximation of the Kirchhoff-Helmholtz Representation of Scattered Fields," *IEEE Transactions on Antennas and Propagation*, Vol. AP-23, No. 5, July 1975, pp. 864–876.

[70]Gounaud, R. (Senior Trend SPO), Memorandum to Col Skip Anderson (Senior Trend JTF), Norm Nelson (Lockheed ADP), and Lt Col Mike Marks (RDPJ), "First Flight Release of Aircraft 781," DACOM No. 003584, Sept. 23, 1981.

[71]Gounaud, R. (Senior Trend SPO), Memorandum to Lt Col Mike Marks (U.S. Air Force/RDPJ), "Technical Problems with Detection, Recognition, Bearing and Ranging," July 23, 1982.

[72]Greenhous, B., "The Israeli Experience," *Case Studies in the Achievement of Air Superiority*, edited by B. F. Cooling, Center for Air Force History, Washington, DC, 1994.

[73]Grier, P., "The (Tacit) Blue Whale," *Air Force Magazine*, Vol. 79, No. 8, Aug. 1996, pp. 50-55.

[74]Gupta, I. J., and Burnside, W. D., "A Physical Optics Correction for Backscattering from Curved Surfaces," *IEEE Transactions on Antennas and Propagation*, Vol. AP-35 No. 5, May 1987.

[75]Haggard, Maj, U.S. Air Force/RDQ-LO, *Background Paper on Lockheed Corporation Technology*, Jan. 7, 1985.

[76]Hall, R. C., "Early Spacecraft Technology," Technology and the Air Force: A Retrospective Assessment, Andrews AFB, Oct. 23–24, 1995.

[77]Hallion, R. P., *The Gulf Air War: A Survey*, Office of the Air Force Historian, Washington, DC (undated).

[78]Headquarters, U.S. Air Force, Deputy Chief of Staff for Operations, *System Operational Concept for the Advanced Technology Aircraft (ATA)*, May 5, 1983.

[79]Headquarters, U.S. Air Force, Deputy Chief of Staff for Operations, *System Operational Concept for the Advanced Technology Aircraft (ATA)*, March 1985.

[80]Headquarters, U.S. Air Force, Deputy Chief of Staff for Operations, *System Operational Concept for the Advanced Technology Aircraft (ATA)*, Feb. 1987.

[81]Headquarters, U.S. Air Force, Deputy Chief of Staff for Research and Development (HQ U.S. Air Force/RD), Memorandum to AFSC/CV, April 21, 1977.

[82]Headquarters, U.S. Air Force/RD, *Advanced Technology Aircraft (ATA) Funding Summary*, undated briefing chart, context indicates 1978.

[83]Headquarters, U.S. Air Force/RD, *Point Paper on ... Funds Transfer*, undated, with attachment, "Senior Trend budget sheets," dated June 9, 1981.

[84]Headquarters, U.S. Air Force/RD, *Senior Trend Congressional Descriptive Summary*, (briefing charts), Aug. 19, 1981.

[85]Headquarters, U.S. Air Force/RD, *[Low Observables Programs] Briefing*, Aug. 28, 1981.

[86]Headquarters, U.S. Air Force/RD, *Acquisition Approach* for Senior Trend Airframe and Engine Support, undated but applies to Oct. 1, 1983 through Sept. 30, 1984.

[87]Headquarters, U.S. Air Force/RD, *F-117A Case Study*, briefing to Deputy Secretary of Defense, April 3, 1991.

[88]Headquarters, U.S. Air Force/RD, *Have Blue*, undated briefing.

[89]Headquarters, U.S. Air Force, Deputy Chief of Staff for Research and Development (low observables) (U.S. Air Force/RDPJ), untitled and undated Point Paper with three attachments: (1) *Funding Alternatives for FY79 Acceleration*, June 8, 1978 with enclosure, *[Low Observables] Short-Term Funding Requirements*, June 7, 1978; (2) *[Low Observables] Program Decision Memorandum Add-On Requirements*, June 8, 1978; and (3) *[Low Observables] Program*, June 6, 1978.

[90]Headquarters, U.S. Air Force/RDPJ, *FY81 Budget Submission*, briefing, undated, context indicates FY80.

[91]Headquarters, U.S. Air Force/RDPJ, *Full-scale Development Program Funding Estimates*, briefing chart, Sept. 11, 1980.

[92]Headquarters, U.S. Air Force/RDPJ, *FY82 Budget Issues*, briefing charts, March 1981.

[93]Headquarters, U.S. Air Force/RDPJ, *Flight Test Accomplishments*, briefing charts, Dec. 8, 1981.

[94]Headquarters, U.S. Air Force/RDPJ, *Senior Trend Budget Assumptions*, briefing charts, April 8, 1982.

[95]Headquarters, U.S. Air Force/RDPJ, *Senior Trend Budget Exercise*, undated briefing charts, approximately April 1982.

[96]Headquarters, U.S. Air Force/RDPJ, *Senior Trend Accident Report*, June 7, 1982.

[97]Headquarters, U.S. Air Force/RDPJ, *Business Approach: Airframe*, undated briefing chart.

[98]Headquarters, U.S. Air Force/RDPJ, *Business Approach: Engine*, undated briefing chart.

[99]Heppe, R.R., (Lockheed) Letter to Dr. R. De Lauer (USD R&E), "Senior Trend Program Report," Aug. 2, 1984.

[100]Herman, R. (assistant SECAF), Memorandum to SECAF, Subject: Senior Trend additional FY82 procurement, June 15, 1981.

[101]Hickey, H., U.S. Air Force Aeronautical Systems Center, Wright–Patterson AFB, Comments provided to the authors, April 1996.

[102]Hillaker, H. J., *Useful Insights into the YF-16 Prototype Program Approach*, Draft, The RAND Corp., 1995.

[103]Hillaker, H. J., Interview with Albert Piccirillo, Aug. 1995.

[104]Jane's Information Group, *Jane's All The World's Aircraft*, 1971–2 through 1995–6 editions.

[105]Johnson, C. L., Lockheed Aircraft Company, "Reduction of Radar Cross Section of Large High Altitude Aircraft," *Proceedings of the 1975 Radar Camouflage Symposium*, Air Force Avionics Laboratory Technical Rept. AFAL-TR-75-100, Dec. 1975.

[106]Johnson, C. L., Lockheed Aircraft Company, "Prototype Experience in the Skunk Works," *Aircraft Prototype and Technology Demonstrator Symposium*, AIAA, New York, 1983.

[107]Kaminski, P. G., Col, U.S. Air Force/RDPJ, Memorandum for the Record, "Coordination with Congressional Committees relative to transfer...to Senior Trend in FY81," July 20, 1981.

[108]Kaminski, P. G., Col, U.S. Air Force/RDPJ, Memorandum for the Record, "Coordination with Congressional Committees relative to transfer...to Senior Trend in FY81," July 23, 1981.

[109]Kaminski, P. G., Col, U.S. Air Force/RDPJ, Memorandum for the Record, "Coordination with Congressional Committees relative to transfer...to Senior Trend in FY81," July 24, 1981.

[110]Kaminski, P. G., Col, U.S. Air Force/RDPJ, Staff Summary Sheet, "[Low Observables] FY83 Funding Requirements," Aug. 19, 1981.

[111]Kaminski, P. G., Col, U.S. Air Force/RDPJ, Staff Summary Sheet, "Senior Trend Procurement Strategy," May 21,1982, with attachment, *Point Paper on Senior Trend Procurement*.

[112]Kaminski, P. G., Col, U.S. Air Force/RDPJ, Memorandum to Col Pete Winters (test site), "Senior Trend Test Problems," DACOM No. 09-126, Sept. 22, 1982.

[113]Kaminski, P. G., Col, U.S. Air Force/RDPJ, Staff Summary Sheet, "Determination and Findings—Advanced Technology Aircraft," Nov. 12, 1982.

[114]Kaminski, P. G., under secretary of defense (acquisition and technology), "The Defense Acquisition Challenge: Technological Supremacy at an Affordable Cost," National Security Industrial Association JAST Industry Conference, Jan. 31, 1995.

[115]Kaminski, P. G., under secretary of defense (acquisition and technology), "Low Observables," Technology and the Air Force: A Retrospective Assessment, Andrews AFB, Oct. 23–24, 1995.

[116]Kaminski, P.G., under secretary of defense (acquisition and technology), Meeting/Discussion of F-117A experience, Feb. 2, 1996.

[117]Keller, J. B., "Geometrical Theory of Diffraction," *Journal of the Optical Society of America*, Vol. 52, No. 2, Feb. 1962.

[118]Kelly, J. D., and Taylor, G. A., *Radar Cross Section Design Handbook Preliminary Data and Format*, Air Force Flight Dynamics Laboratory Report No. AFFDL-TR-74-87, Boeing Aerospace Co., Seattle, WA, Oct. 1974.

[119]Kelly, J. D., Boeing Aerospace Co., "Configuration Design of Low RCS," *Proceedings of the 1975 Radar Camouflage Symposium*, Air Force Avionics Laboratory Technical Rept. AFAL-TR-75-100, Dec. 1975.

[120]Knaack, M. S., *Encyclopedia of U.S. Air Force Aircraft and Missile Systems, Volume 1: Post-World War II Fighters*, Office of Air Force History, Washington, DC, 1978.

[121]Knauth, P., Lt Col, U.S. Air Force (Ret), Comments provided to the authors, Feb. 1996.

[122]Knott, E. F., and Senior, T. B. A., *Non-Specular Radar Cross Section Study*, Air Force Avionics Laboratory Rept. AFAL-TR-73-2, Univ. of Michigan Radiation Laboratory, Ann Arbor, MI, Feb. 1973.

[123]Knott, E. F., Shaeffer, J., and Tuley, M., *Radar Cross Section*, Artech House, Norwood, MA, 1985.

[124]Lockheed Advanced Development Co., *The Skunk Works Approach to Aircraft Development, Production and Support*, Sunland, CA, Aug. 1992.

[125]Lockheed Advanced Development Projects (ADP), Advanced Technology Aircraft (ATA) Budget and Schedule Summary, Lockheed Rept. SP4763F, Feb. 27, 1978, with cover letter from Ben Rich (Lockheed ADP) to Lt Col Ken Staten (RDPJ).

[126]Lockheed ADP, *Have Blue Programmatic Review*, Sept. 5, 1979.

[127]Lockheed ADP, *Two Hour Briefing*, April 11, 1984.

[128]Lockheed ADP, *F-117A Cost Performance and Contracts History*, DCN TA189-003112, April 19, 1989, with cover memorandum from Col Mike Marks, SAF/AQL, May 3, 1989.

[129]Lockheed Martin Corp. World Wide Web Homepage (www.lockheed.com).

[130]Lockheed/U.S. Air Force, *Senior Trend Rate Production Study*, Dec. 1982.

[131]"Lockheed's Kelly Johnson Building 'Stealth' Aircraft," *Aerospace Daily*, July 23, 1976, pp. 121, 122.

[132]Locus, S. S., Heath, H. C., Coleman, J. R., and Mitzner, K. M., *Calculation of Radar Cross Section, Vol. I: User's Guide*, Air Force Avionics Laboratory Rept. AFAL-

TR-74-112, Vol. I, Northrop Corp., Hawthorne, CA, Sept. 1974.

[133]Lorber, H. W., Wintersdorf, R. W., and Cota, G. R., *Low RCS Vehicle Study*, Air Force Avionics Laboratory Rept. AFAL-TR-74-320, Teledyne Ryan Aeronautical, San Diego, CA, Jan. 31, 1975.

[134]Lorrain, P., and Corson, D., *Electromagnetic Fields and Waves* (2nd Ed.), W. H. Freeman and Co., San Francisco, 1962.

[135]*Low Observables Strategic Vision Assessment Panel*, Meeting and discussion, Arlington, VA, July 30, 1996.

[136]Madden, T. L., AFAL/WRP, "Aircraft Applications," *Proceedings of the 1975 Radar Camouflage Symposium*, Air Force Avionics Laboratory Technical Rept. AFAL-TR-75-100, Dec. 1975.

[137]*Management Fundamentals for Systems Acquisition Course*, H. J. Ford Associates Inc., Arlington, VA, 1995.

[138]Marks, M., Lt Col, U.S. Air Force/RDPJ, Memorandum to Col Don Merkl (Senior Trend SPO) and Col Skip Anderson (Senior Trend JTF), Subject: Film, May 8, 1981.

[139]Marks, M., Lt Col, U.S. Air Force/RDPJ, Memorandum to Scofield/Trembour/Vogan, "FY83 President's Budget," Dec. 10, 1981.

[140]Marks, M., Lt Col, U.S. Air Force/RDPJ, Memorandum to Col Richard Scofield (Senior Trend SPO), "FY82 Reprogramming," March 11, 1982.

[141]Marks, M., Lt Col, U.S. Air Force/RDPJ, Memorandum to Col Richard Scofield, "Senior Trend Improvements Plan," April 8, 1982.

[142]Marks, M., Lt Col, U.S. Air Force/RDPJ, Memorandum to Skip Hickey (Senior Trend SPO) and Lt Col John Madia, "F-117 vs. SU AWACS," DACOM No. 07-077, July 21, 1982.

[143]Marks, M., Lt Col, U.S. Air Force/RDPJ, Memorandum to Col Dick Scofield (Senior Trend SPO), "Senior Trend Test Problems," DACOM No. 09-107, Sept. 20, 1982.

[144]Marks, M., Lt Col, U.S. Air Force/RDPJ, Memorandum to Scofield/Trembour/Vogan, "Senior Trend Procurement Plans," undated.

[145]Marks, M., Lt Col, U.S. Air Force/RDPJ, *Point Paper on Senior Trend Restructured Procurement Plan*, Nov. 3, 1981.

[146]Marks, M., Lt Col, U.S. Air Force/RDPJ, *Point Paper on Senior Trend IOC Date*, May 12, 1982.

[147]Marks, M., Lt Col, U.S. Air Force/RDPJ, *Point Paper on Senior Trend Program Budget Decision (PBD)*, Nov. 23, 1982.

[148]Marks, M., Lt Col, U.S. Air Force/RDPJ, *Point Paper on Senior Trend Program Summary*, Nov. 23, 1982.

[149]Marks, M. D., Col, U.S. Air Force (Ret), Comments provided to the authors, July 1996.

[150]Marsh, R. T., Gen, U.S. Air Force, Memorandum to Gen Charles A. Gabriel, Subject: Tailpipe Cracks, July 1, 1983.

[151]Martin, E., Comments provided to the authors, Aug. 1996.

[152]Mathis, R. C., Lt Gen, U.S. Air Force, AFSC/CV, Memorandum to U.S. Air Force Headquarters/RD, May 18, 1977.

[153]McMullen, T. H., Lt Gen, U.S. Air Force (TAC/CV), Memorandum to Lt Gen Kelly H. Burke (U.S. Air Force Headquarters/RD), "Senior Trend Procurement Strategy," June 26, 1982.

[154]Mentzer, C. A., AFAL/WRP, "Non-Specular RCS Reduction," in *Proceedings of the 1975 Radar Camouflage Symposium*, Air Force Avionics Laboratory Technical Rept. AFAL-TR-75-100, Dec. 1975.

[155]Merkl, E. D., Col, U.S. Air Force (Senior Trend SPO), Memorandum to Lt Col Ken Staten (RDPJ), "Senior Trend Budget," Aug. 15, 1979.

[156]Merkl, E. D., Col, U.S. Air Force (Senior Trend SPO), Memorandum to Col Ken Staten (RDPJ), "What-If Exercise," Oct. 3, 1980.

[157]Merkl, E. D., Col, U.S. Air Force (Senior Trend SPO), Memorandum to Col Ken Staten (RDPJ), "Per Unit Senior Trend Aircraft Flyaway Costs," undated but in response to inquiry of Nov. 25, 1980.

[158]Merkl, E. D., Col U.S. Air Force (Senior Trend SPO), Memorandum to Col Ken Staten (RDPJ), Subject: long-lead funding, March 4, 1981.

[159]Merkl, E. D., Col, U.S. Air Force (Senior Trend SPO), Memorandum to Alan Brown (Lockheed ADP), Col Ken Staten (RDPJ), R.K. Bertelson (General Electric), Col Skip Anderson (Senior Trend JTF), and Col Bob Jackson (4450th Tactical Group), Subject: First Flight Readiness Review, March 6, 1981.

[160]Merkl, E. D., Col, U.S. Air Force (Senior Trend SPO), Memorandum to Alan Brown (Lockheed ADP), Col Ken Staten (RDPJ), R. K. Bertelson (General Electric), Col Skip Anderson (Senior Trend JTF), and Col Bob Jackson (4450th Tactical Group), "First Flight Readiness Review Scheduled," undated, approximately March 15, 1981.

[161]Merkl, E. D., Col, U.S. Air Force (Senior Trend SPO), Memorandum to Col Ken Staten (RDPJ), DACOM No. 000611, Subject: First Flight Reporting Procedures, March 17, 1981.

[162]Merkl, E. D., Col U.S. Air Force (Senior Trend SPO), Memorandum to Col Ken Staten (RDPJ), Subject: long-lead funding, March 23, 1981.

[163]Merkl, E. D., Col, U.S. Air Force (Senior Trend SPO), Memorandum to Norm Nelson (Lockheed ADP), Col Ken Staten (RDPJ), Col Skip Anderson (Senior Trend JTF), Col Bob Jackson (4450th Tactical Group), Subject: First Flight Readiness Review, March 26, 1981.

[164]Merkl, E. D., Col, U.S. Air Force (Senior Trend SPO), Memorandum to Alan Brown (Lockheed ADP), Col Ken Staten (RDPJ), R.K. Bertelson (General Electric), Col Skip Anderson (Senior Trend JTF), Col Bob Jackson (4450th Tactical Group), April 2, 1981.

[165]Merkl, E. D., Col, U.S. Air Force (Senior Trend SPO), Memorandum to Col Ken Staten (RDPJ), Col Bob Jackson (4450th Tactical Group), Col Skip Anderson (Senior Trend JTF), April 10, 1981.

[166]Merkl, E. D., Col, U.S. Air Force (Senior Trend SPO), Memorandum to Col Ken Staten (RDPJ), Col Bob Jackson (4450th Tactical Group), and Col Skip Anderson (Senior Trend JTF), "Status Report for A/C No. 780," DACOM No. 001137, April 17, 1981.

[167]Merkl, E. D., Col, U.S. Air Force (Senior Trend SPO), Memorandum to Col Ken Staten (RDPJ), Col Bob Jackson (4450th Tactical Group), Col Skip Anderson (Senior Trend JTF), April 22, 1981.

[168]Merkl, E. D., Col, U.S. Air Force (Senior Trend SPO), Memorandum to Col Ken Staten (RDPJ), April 29, 1981.

[169]Merkl, E. D., Col, U.S. Air Force (Senior Trend SPO), Memorandum to Norm Nelson (Lockheed ADP), Col Ken Staten (RDPJ), Col Bob Jackson (4450th TG), Col Skip Anderson (Senior Trend JTF), "Minutes of First Flight Readiness Review 27-8-9, April 1981 Including Post-Fire Session," April 30, 1981.

[170]Merkl, E. D., Col, U.S. Air Force, and Baber, J. W., Lt Col, U.S. Air Force, (Senior Trend SPO) Memorandum to Norm Nelson (Lockheed ADP) and Lt Col Mike Marks (RDPJ), "Senior Trend Lot II Production Delivery Schedule," DACOM No. 002576, July 21, 1981.

[171]Merritt, M., Lt Col, SAF/AQL, *Background Paper on F-117A Deliveries*, 21 May 1990.

[172]Merritt, M., Lt Col, SAF/AQL, *Background Paper on Have Blue*, Feb. 25, 1991.

[173]Michaeli, A., *Equivalent Edge Currents for Arbitrary Aspects of Observation, IEEE Transactions on Antennas and Propagation*, Vol. AP-32 No. 3, March 1984.

[174]Miller, B., "Advanced Tactical Fighter Studies Set," *Aviation Week and Space Technology*, Jan. 19, 1976, pp. 16-17.

[175]Miller, J., *Lockheed's Skunk Works: The First Fifty Years*, Aerofax, Inc., Arlington, TX, 1993.

[176]Mitzner, K. M., *Incremental Length Diffraction Coefficients*, Northrop Corp. Aircraft Division, Technical Rept. AFAL-TR-73-296, April 1974.

[177]Mitzner, K. M., *Studies in Physical Optics*, Northrop Corporation Aircraft Division, Air Force Avionics Laboratory Technical Rept. AFAL-TR-74-22, May 1974.

[178]"More Sources Report Downed Plane Stealth," Associated Press release, Bakersfield, CA, July 13, 1986.

[179]Moseley, R., Lt Col, U.S. Air Force, *Senior Trend Test Program from Concept to Initial Operational Capability: Planned vs. Actual*, Briefing, Dec. 20, 1983 with cover memo from Col Michael E. Sexton to Col Paul G. Kaminski dated Dec. 21, 1983.

[180]Mullin, S. N., (president, Lockheed Advanced Development Company), "The Evolution of the F-22 Advanced Tactical Fighter," Wright Brothers Lecture, AIAA Aircraft Design Systems Meeting, Aug. 24, 1992.

[181]Mullin, S. N. (Lockheed program manager for the F-117A from Feb. 1982 to March 1985), Letter to David C. Aronstein, Aug. 17, 1996.

[182]O'Malley, J. F., Gen, U.S. Air Force, vice chief of staff, Memorandum for AF/CC, "Senior Trend Procurement Strategy," July 7, 1982.

[183]Office of the Assistant Secretary of the Air Force, Directorate of Special Programs, Memorandum to Col Hoh and Col Marks, "Senior Trend Specifications," signature illegible, Nov. 10, 1988.

[184]Office of the Secretary of Defense (Comptroller), miscellaneous program budget decisions (PBDs), Nov.–Dec. 1982.

[185]Oshiro, F. K., Mitzner. K. M., Locus, S.S., Coleman, J. R., and Heath, H. C., *Calculation of Radar Cross Section*, Part II (Analytical Report) and Part III (User's Manual), Northrop Corp., Air Force Avionics Laboratory Rept. No. AFAL-TR-70-21, April 1970.

[186]Overholser, D. D., Comments provided to the authors, Aug. 2, Aug. 21, and Aug. 29, 1996.

[187]Pace, S., *Lockheed Skunk Works*, Motorbooks International, Osceola, WI, 1992.

[188]Perko, K., Comments provided to the authors, Sept. 1996.

[189]Piccirillo, A. C., "The Have Blue Technology Demonstrator and Radar Cross Section Reduction," AIAA/SAE World Aviation Congress (Paper No. 965538), Los Angeles, Oct. 1996.

[190]Pinney, C., Lt Col, U.S. Air Force, Joint Strike Fighter Avionics IPT Lead (formerly with the Senior Trend CTF), Comments provided to the authors, March 1996.

[191]Price, A., *Focke-Wulf 190 At War*, Scribner's, New York, 1977.

[192]Przemieniecki, J. S., Ed., *Acquisition of Defense Systems*, AIAA Education Series, AIAA, Washington, DC, 1993.

[193]Puckett, A. E., and Stewart, H. J., *Journal of the Aeronautical Sciences*, Vol. 14, pp. 567-578, 1947.

[194]Reiman, J. A., U.S. Air Force Aeronautical Systems Center, Wright–Patterson AFB, Comments provided to the authors, March–April 1996.

[195]Rich, B. (Lockheed ADP), Letter to Lt Col Ken Staten, "Senior Trend Updated Budgetary Estimates," Nov. 1, 1978.

[196]Rich, B., *The Skunk Works Approach*, briefing to Defense Science Board, Air Force Academy, Colorado Springs, CO, July 25, 1990.

[197]Rich, B. R., and Janos, L., *Skunk Works: A Personal Memoir of My Years at Lockheed*, Little, Brown and Co., New York, 1994.

[198]Ridenour, L. N., Ed., *Radar System Engineering*, Massachusetts Institute of Technology Radiation Laboratory Series, McGraw-Hill, New York, 1947.

[199]Robinson, C. A., Jr., "Multipurpose Bomber Advances," *Aviation Week and Space Technology*, Aug. 4, 1980.

[200]Rublee, G., and Sawyer, M., *Global Power Items of Interest*, ANSER Congressional Analysis Team, March 26, 1996.

[201]Russ, R. D., Maj Gen, director of operational requirements, DCS RD&A, U.S. Air Force headquarters, Memorandum to General Mathis, AF/CV, "Additional four Senior Trend aircraft for FY82: Action Memorandum," May 15,1981.

[202]Scherrer, Norrup, Jusko, and Kong, *Radar Attenuating Material (RAM) Composite Test Program*, Lockheed-California Co., Burbank, CA, April 1969.

[203]Scherrer, R., Overholser, D. D., and Watson, K. E., Patent Application for "Vehicle," Docket No. P-01-1532, Lockheed Corp., Burbank, CA, Jan. 16, 1979.

[204]Scherrer, R., Comments provided to the authors, Aug. 1996.

[205]Scofield, R. M., Col, U.S. Air Force (Senior Trend SPO) Memorandum to Norm Nelson (Lockheed ADP), Col Skip Anderson (Senior Trend JTF), and Lt Col Mike Marks (RDPJ), "First Flight Release of Aircraft 781," DACOM No. 003268, Sept. 4, 1981.

[206]Scofield, R. M., Col, U.S. Air Force (Senior Trend SPO), Memorandum to Lt Col Mike Marks and Lt Col John Madia (RDPJ), "Program Funding Requirement," DACOM No. 004678, Dec. 7, 1981.

[207]Scofield, R. M., Col, U.S. Air Force (Senior Trend SPO), Memorandum to Lt Col Mike Marks and Lt Col John Madia (RDPJ), "Program Direction and Funding," Jan. 4, 1982.

[208]Scofield, R. M., Col, U.S. Air Force (Senior Trend SPO), Memorandum to Lt Col John Madia (RDPJ), "Senior Trend Funding Alternatives," April 20, 1982.

[209]Scofield, R. M., Col, U.S. Air Force (Senior Trend SPO), Memorandum to Lt Col Mike Marks (RDPJ), "Senior Trend Test Problems," DACOM No. 9428, Sept. 22, 1982.

[210]Senior Trend Joint Test Force (JTF), "First Flight Preliminary Report," June 18, 1981.

[211]Senior Trend Joint Test Force (JTF), *Flight Test Progress Report for Senior Trend Aircraft, Report No. 10*, July 1982.

[212]Siefke, S. P., Lt Col, U.S. Air Force, chief, Weapon Flight Mechanics Division, Armament Directorate, Wright Laboratory, Comments provided to the authors, July 3, 1996.

[213]Skolnik, M., *Radar Handbook* (Second Edition), McGraw-Hill, New York, 1990.

[214]Smith, G., *Application of F-117 Acquisition Strategy to Conventional Programs in the New Acquisition Environment*, Briefing, The RAND Corp., Control no. RAN/ST-072/94, 1994.

[215]Smith, G., *Preliminary Analysis of the F-117 Acquisition Strategy*, Draft, The RAND Corp., Control No. RAN/ST-0780/94, July 1994.

[216]Smith, R. A., Lt Col, U.S. Air Force, Joint Strike Fighter Flight Systems IPT Lead (formerly with the 2762nd Logistics Squadron (Special), Detachment 6, Norton AFB), Comments provided to the authors, March 1996.

[217]Staten, K., Lt Col, U.S. Air Force/RDPJ, Staff Summary Sheet, "Delegation of Special Authority and Determination and Findings," Sept. 30, 1980.

[218]Staten, K., Col, U.S. Air Force/RDPJ, Memorandum to Col Don Merkl (Senior Trend SPO), Subject: Attendance at First Flight, March 3, 1981.

[219]Staten, K., Col U.S. Air Force/RDPJ, Memorandum to Col Don Merkl, Subject: additional four FY82 Senior Trend aircraft, March 19, 1981.

[220]Staten, K., Col, U.S. Air Force/RDPJ, Memorandum to Col Don Merkl, ASD/RWSX (Senior Trend SPO), "Senior Trend Additional Procurement," May 18, 1981.

[221]Staten, K., Col, U.S. Air Force/RDPJ, Memorandum to Col Don Merkl and Lt Col Jerry Baber (Senior Trend SPO), "Funding Profiles," May 22, 1981.

[222]Staten, K., Col., U.S. Air Force/RDPJ, Memorandum for the Record, "Coordination with Congressional Committees relative to transfer... to Senior Trend in FY81," July 25, 1981.

[223]Staten, K., Col U.S. Air Force/RDPJ, Memorandum to Col Don Merkl, "Assignment of Aircraft Serial Numbers," undated.

[224]Swarts, Lt Col, U.S. Air Force/RDPJ, Staff Summary Sheet, "Long-lead Materials, Hardware, and Data...," Nov. 20, 1979.

[225]Sweetman, B., *Stealth Aircraft: Secrets of Future Airpower*, Motorbooks International, Osceola, WI, 1986.

[226]Sweetman, B., "The Stealth Master," *Interavia Aerospace Review*, Feb. 1992.

[227]Sweetman, B., "The Invisible Men" (draft, 1997).

[228]Sweetman, B., "The Invisible Men," *Air and Space*, Vol.12, No.1, 1997, pp.18–27.

[229]Sweetman, B., and Goodall, J., *Lockheed F-117A: Operation and Development of the Stealth Fighter*, Motorbooks International, Osceola, WI, 1990.

[230] "The Oxcart Story," *Air Force Magazine*, Nov. 1994, pp. 40–47.

[231]Trembour, B., Lt Col, U.S. Air Force, Memorandum to Lt Col John Casper, "F-117 vs. SU AWACS," July 22,1982.

[232]Tweed, J., *F-117A Milestones*, Lockheed Martin Skunk Works, Palmdale, CA, Dec. 4, 1990 (revised Aug. 18, 1996).

[233]Twigg, J. K., Col, U.S. Air Force (Ret), *A Program Manager's Views* (briefing), Aug. 30, 1995.

[234]Twigg, J. K., Col, U.S. Air Force (Ret), Letter to Albert C. Piccirillo, Feb. 29, 1996.

[235]Twigg, J. K., Col, U.S. Air Force (Ret), Interview with David C. Aronstein, April 24, 1996.

[236]Twigg, J. L., *"To Fly and Fight": Norms, Institutions, and Fighter Aircraft Procurement in the United States, Russia, and Japan*, Ph.D. dissertation, Department of Political Science, Massachusetts Inst. of Technology, Cambridge, MA, Sept. 1994.

[237]Ufimtsev, P. I., *Method of Edge Waves in the Physical Theory of Diffraction*, Soviet Radio Publication House, Moscow, 1962 (translation by U.S. Air Force Systems Command, Foreign Technology Division, Document No. FTD-HC-23-259-71, 1971).

[238]U.S. Air Force, *Application of Special Management*, AFR 800-29, Feb. 11, 1982 (superseded earlier AFR 66-22).

[239]U.S. Air Force 46th Test Group, *RATSCAT Signature Measurement Short Course Notes*, Holloman AFB, NM, undated.

[240]U.S. Air Force Systems Command, Aeronautical Systems Division, Acquisition Plan No. SD26RX-80-2, *Advanced Technology Aircraft (Senior Trend)*, signed Lt Col Jerry W. Baber (business manager), Col Eldred D. Merkl (program manager), and Col David B. Englund (systems program director), Wright–Patterson AFB, OH, Oct. 31, 1979.

[241]U.S. Air Force Systems Command, Aeronautical Systems Division, Acquisition Plan No. RWSX-80-3, *Advanced Technology Aircraft (Senior Trend)*, signed Lt Col Jerry W. Baber (business manager), Col Eldred D. Merkl (program manager), and Col David B. Englund (systems program director), Wright–Patterson AFB, OH, July 31, 1980.

[242]U.S. Air Force Systems Command, Aeronautical Systems Division, Acquisition Plan No. RWSX-80-3, Amendment No. 1, *Advanced Technology Aircraft (Senior Trend)*, signed Lt Col Jerry W. Baber (business manager), Col Eldred D. Merkl (program manager), and Col David B. Englund (systems program director), Wright–Patterson AFB, OH, April 21, 1981.

[243]U.S. Air Force Systems Command, Aeronautical Systems Division, Acquisition Plan No. XRJ-83-1, Amendment No. 2, *Advanced Technology Aircraft (Senior Trend)*, signed Lt Col Jerry W. Baber (business manager), Col Richard M. Scofield (program manager), and Col John K. Twigg (systems program director), Wright–Patterson AFB, OH, Nov. 3, 1982.

[244]U.S. Air Force Systems Command, Aeronautical Systems Division, *Senior Trend Program Master Schedule*, Rev. C, Aug. 1980.

[245]U.S. Air Force Systems Command, Aeronautical Systems Division, *F-117 Program Summary*, ASD/YSRS Program Integration, undated briefing.

[246]U.S. Air Force, Air Force Systems Command, *USAF Prototype Study,* Final Rept. [no report number assigned], Sept. 1971.

[247]U.S. Air Force Tactical Air Command, *Advanced Technology Aircraft Preliminary Operational Program Introduction Document (PID)*, undated, with cover memorandum from Col Robert A. Jackson, TAC 4450th Tactical Group commander, to U.S. Air Force Headquarters/RDPJ, May 2, 1980.

[248]U.S. Air Force Wright Laboratories, *Advanced Low-Observable Optical Air Data System (ALOADS)*, Briefing, Wright–Patterson AFB, OH, 1995.

[249]Vincenti, Nielson, and Matteson, *Investigation of Wing Characteristics at a Mach Number of 1.53. Part I: Triangular Wings of Aspect Ratio 2*, NACA Research Memorandum A7I10, 1947.

[250]Watson, W. H., *The Method of Spherical Representation for Treating Electromagnetic Scattering*, Lockheed Missiles and Space Co., Sunnyvale, CA, April 1966.

[251]Watson, W. H., and Kusnezov, N., *Electromagnetic Scattering by Regular Polyhedra*, Lockheed Missiles and Space Co., Sunnyvale, CA, April 1967.

[252]Welch, L. D., Maj Gen, U.S. Air Force, DCS Operations, Memorandum to U.S. Air Force Headquarters/RDQ.

[253]Weyermuller, A. P., Col, U.S. Air Force, *Stealth Employment in the Tactical Air Force —A Primer on Its Doctrine and Operational Use*, U.S. Army War College, Feb. 1992.

[254]Wilson, G. C., "50 'Stealth' Fighters in Operation," *Washington Post*, Aug. 22, 1986.

[255]Winters, P., Col, U.S. Air Force (Test Site), Memorandum to Col Dick Scofield, "Senior Trend Test Problems," DACOM number and date unavailable, approximately Sept. 22, 1982.

[256]Wintersdorf, R. W., Teledyne Ryan Aeronautical Company, "RPV - Drone Applications," *Proceedings of the 1975 Radar Camouflage Symposium*, Air Force Avionics Laboratory Technical Rept. AFAL-TR-75-100, Dec. 1975.

[257]Woods, J., Jr., and Rose, J. G., *Quiet Attack Aircraft Radar Cross Section Test Results*, Rept. no. MDC-A2949-2, McDonnell Douglas Corp., Nov. 15, 1974.

[258]Young, J. O., *Kelly's Way*, video documentary produced by the Flight Test Historical Foundation, Edwards AFB, CA, 1993.

Index